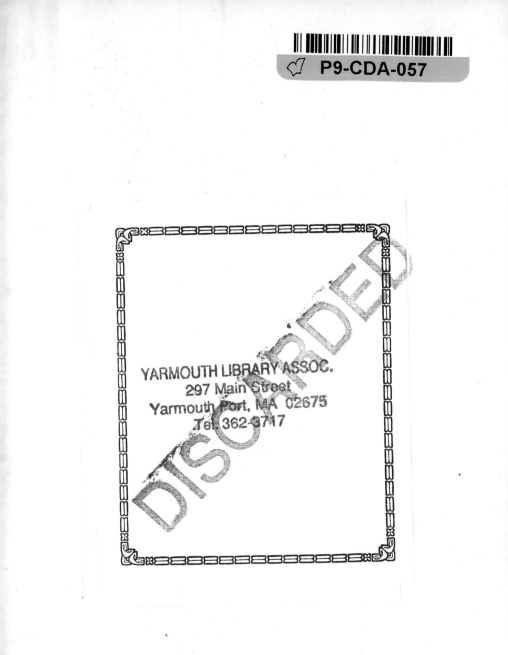

DISCOVERY

Books by Mahlon Hoagland

THE ROOTS OF LIFE: A Layman's Guide to
Genes, Evolution, and the Ways of Cells

DISCOVERY: The Search for DNA's Secrets

DISCOVERY

The Search for DNA's Secrets

———

Mahlon Hoagland

Boston

Houghton Mifflin Company

1981

Library of Congress Cataloging in Publication Data

Hoagland, Mahlon B.
Discovery, the search for DNA's secrets.

Includes index.
1. Genetics — History.   2. Molecular genetics —
History.   3. Deoxyribonucleic acid.   I. Title.
QH428.H63   574.87′3282   81-6560
ISBN 0-395-30510-1   AACR2

Printed in the United States of America

P 10 9 8 7 6 5 4 3 2 1

For my father

HUDSON HOAGLAND

Contents

viii   *Contents*

molecular structure. A chronicle of James D.
Watson and Francis Crick's revelation of DNA
structure.

Darwinian natural selection and the intimate relatedness of all living forms are confirmed by
molecular biology. The ways of science are
unique in revealing verifiable truth. New realms
of exploration stretch before us since the cracking of the code.

Preface

THIS ALL BEGAN with an oyster. I was perched on my favorite stool before the oyster bar at the Union Oyster House in Boston preparing to dispatch six of the tasty bivalves when it suddenly hit me how old they were. Here I was about to nourish myself on a creature that had remained essentially unchanged for a few hundred million years! Fork poised motionless, elbows straddling my plate, I pondered this moving thought. They were in a sense immortal, weren't they? At least, they were immortal enough for me. The instructions for making these oysters, written in their DNA, had been passed along from generation to generation of oysters with little change for hundreds and hundreds of millions of years. One hundred million years is 4 million people generations; it's 50,000 reruns of the Christian era! And for unnumbered more millions of years before the oyster appeared, evolution had been hunt-and-pecking out instructions to produce oysters.

The oyster's staunch persistence along evolution's trail is a marvel, no question about it. Ruminating on this (with fading appetite) led me to a greater marvel: another of earth's creatures reared by the same forces of evolution, the human who consumes the oyster, though only a million or so years old, had been able to discover the oyster's history and, moreover, explain the oyster's success. In fact, humans had been able to pry from nature the general rules of inheritance and the grand procession of evolution of *all* creatures. And they had accomplished all that in only about *one hundred years* of exploration!

What other achievement of man is even remotely comparable? How had it been done? How had the human animal's burning need to know been disciplined and focused and energized sufficiently to transform all that mystery into enlightenment? Where did that penetrating power of discovery come from?

It then hit me forcibly that *I knew*. At least I knew as much as anyone knew. As a scientist, I was reasonably expert on such matters, and I took it all for granted. I knew how scientists work, exploring by careful observation, imagining explanations, and testing those explanations by experiment to see whether they are right. I knew that scientists' experiments are prophecies, predictions of what would happen if they perturbed or provoked nature. I knew that when scientists are on the right track, their theories or predictions are borne out by having the system do what they expected. Even if the scientists' initial guesses are wrong, the unexpected event often leads to a profound new insight. The ingenious management of surprise is a mark of a good scientist. Science's astonishing power of penetrating mystery by means of a simple mode of exploration is endowed with a final clinching effectiveness: no truth produced by one scientist has status until it is validated by someone else. The truths of science are everywhere acceptable because they are everywhere verifiable. Each small truth serves as an essential part of the ever-enlarging construction of knowledge of nature; the whole is unshakable because every element in its assembly in unshakable.

It is this *how* of science that makes a clean, clear story — a story that has really not been told. People know of the products of science, but they know little of the process of science.

Behind the how, of course, is the *why*. Why do we question, probe, search, and explore with such a passion? It seems too easy to say we do it because we're endowed with curiosity, imagination, a need to conquer mystery, or a competitive urge to succeed. These qualities also kindle artists and, indeed, all of us to varying degrees. They bind all of us together in reaching out for something better.

Scientists are the explorers of the twentieth century. Until now, they have largely failed to step across the language barrier to convey the excitement of the search to their nonscientific brothers and sisters. It is as though the accounts of the exploits of Columbus, Magellan, Cook, Drake, and all the rest were locked in a vault, or that those worthies had decided to keep their discoveries a secret. The essence of scientific exploration and discovery can be made as accessible to the motivated nonscientist as can the exploits of the other great explorers of the past.

In recent years scientists have begun to share their experiences with a wider audience of nonscientists who search for a deeper understanding of the natural world and their relationship to it. I believe this personal interest in science's ways and discoveries will surely grow in the coming years. Although one can live one's life believing that the world is flat, particularly when everyone believes it, one's life is fuller for knowing the truth, with all its enriching implications.

❖

Suddenly, with an upsurge of missionary zeal, my appetite returned. I consumed the oysters without guilt, and determined then and there to tell science's most incredible story: the discovery of the gene and how it works.

✻

Now that I've finished the writing, and before your reading of it begins, I want to express admiration for my colleagues and friends Francis Crick, Bernard Davis, Arthur Kornberg, Seymour Benzer, James Watson, François Jacob, and Jacques Monod (now gone) for their lustrous performances. I thank my secretary, Jacqueline Foss, for her skill and good cheer in repeatedly typing the manuscript. And to my wife, Olley Jones Hoagland, who sustains me in the belief that scientists can tell their story to people, I give very special thanks.

MAHLON HOAGLAND
Worcester Foundation for
Experimental Biology
Shrewsbury, Massachusetts, 1981

DISCOVERY

CHAPTER ONE

The Curtain Rises on Modern Genetics

Gregor Mendel starts the science of genetics. Thomas Hunt Morgan locates genes in chromosomes. Hermann Muller makes mutations with x-rays. Are genes made of protein or DNA? The discovery of DNA by Friedrich Miescher. Its basic chemistry. Frederick Griffith discovers bacterial transformation.

THE CURTAIN ROSE on biology's grand new understanding of the gene in 1944. That was the year that disclosed convincing new evidence that genes, the controlling elements of heredity, were made of *deoxyribonucleic acid,* or DNA. Within 20 years science opened up a whole new brilliant world inside the cell, revealing how genes controlled cells and cell inheritance. That tale of discovery is our story.

The determination to discover the chemical nature of the factors that controlled inheritance had, by 1944, become a matter of intense preoccupation among biologists. The science of genetics is rooted in our fascination with how our appearances, traits, propensities, competencies, and maladies are transmitted from parent to child. Genetic exploration has also been promoted by our desire to improve plants and animals for agricultural purposes. Genetics is the study of the modes of inheritance of specific individual qualities or traits, and of how cells package and encode the information needed to insure the onward transmission of individuality.

Gregor Mendel built the basis of modern genetics in the mid-1800s. He was a priest, the only son of a peasant family in Moravia, then a part of Austria (now of Czechoslovakia). He entered a monastery in 1843, apparently as a way of continuing his studies in genetics. Moravia, an agricultural region, had a very practical interest in the breeding of animals and plants.

Mendel did most of his work with pea plants. He chose peas because he could cultivate them easily, they readily fertilized each other, and they grew relatively rapidly. In addition, peas have easily observable and reproducible traits, such as height, location of the flower, color and shape of the pod, and shape of the pea. The essence of Mendel's method was to "mate" purebred pea plants and observe the offspring. He selected two sets of pea plants. Each set of plants had been inbred for many generations for a distinctive trait such as wrinkled or smooth peas, so they were genetically pure for that particular trait. He artificially cross-fertilized these plants. The first-generation offspring were all identical: all had smooth peas. Parents with smooth peas, crossed with parents with wrinkled peas, produced "children" with only smooth peas! What had happened to the hereditary factor that controls the trait of "wrinkledness"? Had it been lost, discarded, or destroyed during reproduction? No, because when the members of this first generation of plants were allowed to fertilize each other, some of *their* offspring had wrinkled peas! So the factor determining wrinkledness was still in the plants, apparently in hiding. Mendel measured the proportions of plants with smooth peas and with wrinkled peas in that second generation; roughly 75 percent were smooth, and 25 percent were wrinkled.

Mendel saw clearly that whatever the heredity-determining factors were in his plants — it was some time later that they acquired the name "genes" — they were discrete, independent entities, that is, real *things*. He hadn't a clue as to what a gene was, of course, but clearly a gene-determined trait could disappear in the first offspring of a mating because its gene was not *expressing* itself. The gene could re-appear, or re-express the trait it controlled, in a second generation. Mendel put it all together this way:

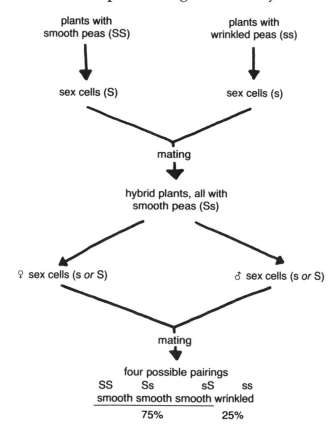

plants with
smooth peas (SS)

plants with
wrinkled peas (ss)

sex cells (S)

sex cells (s)

mating

hybrid plants, all with
smooth peas (Ss)

♀ sex cells (s *or* S)

♂ sex cells (s *or* S)

mating

four possible pairings
SS   Ss   sS   ss
smooth smooth smooth wrinkled
75%   25%

There are two factors (we'll call them by their right name, *genes*) that control each trait of an organism; one is derived from the mother, and one is derived from the father. In the diagram, the shape of peas is governed by the two genes S and s, for smoothness and wrinkledness, respectively.

When plants make their sex cells (pollen and ova) in preparation for mating, each sex cell gets *only one* of the pair of genes, either S or s. When male and female sex cells unite to create a new plant, all of the new plants will contain S paired with s, because that is the only possible combination of genes. These are called hybrid plants. Mendel now postulated that one form of the pea shape-determining gene was *dominant* (S) and one was *recessive* (s). Because gene S is dominant over s, all the hybrid plants' peas will be smooth. When the first-generation hybrid plants in turn produce their sex cells, each pollen grain will be *either* s or S, and each ovum will be *either* s or S. These genes pair up to form the next generation, and if there are enough plants to get good statistics, there will be a 25 percent chance of getting SS progeny, a 25 percent chance of getting Ss, a 25 percent chance of getting sS, and a 25 percent chance of getting ss. Plants containing the SS gene pair will have smooth peas. The hybrid sS and Ss gene pairs will also produce smooth peas, because S is dominant. Only the ss pair will produce wrinkled peas. That adds up to 75 percent smooth and 25 percent wrinkled peas. So Mendel could conclude that the factors governing heredity in plants were able to pair up freely as independent agents and that some were dominant relative to others when paired.

Mendel published his work in 1866. It received little notice in the scientific world and proceeded, incredibly, to sink into obscurity for nearly 50 years! One reason may have been the statistical nature of his proof; in those days, biologists weren't used to statistics. Another was the relative intellectual isolation of scientists at that time; scientists were much more sparsely distributed than they are now.

Thomas Hunt Morgan was the next major figure in genetics. He pushed wide open the door Mendel had unlocked. A professor at Columbia during the first quarter of the 1900s, and then at the California Institute of Technology until his death in 1945, he lived and breathed genetics and passed an exuberant enthusiasm along to his colleagues and students. He believed deeply that genes could be shaken free of ancient mystery, and he drove himself and his students relentlessly in the search.

Morgan chose *Drosophila melanogaster* (the common fruit fly) as his experimental subject. Almost every square inch of space in his labs was covered with stoppered milk bottles humming with fruit flies. These flies could be bred easily, were cheap, and took up little space compared to animals or pea plants. In addition, they had only four chromosomes. (By comparison, we humans have 46.) By the time Morgan began his work, geneticists knew that chromosomes were important to heredity. Chromosomes were structures that could be seen with the aid of the microscope in the nucleus of cells; each one separated into two parts when cells were about to divide, and after cell division they were equally distributed among the daughter cells. (Chromosomes also contained the cell's DNA, a fact not then known.)

Morgan's elegant experimental work showed that genes were on or in the chromosomes. Mendel's indivisible, independently mixing trait-determining things were made part of visible entities in cells, the chromosomes. More than that,

Morgan showed that genes must be arranged *in a linear order* on chromosomes, like a string of beads.

How does one conclude that genes are ordered like a string of beads if one doesn't know what genes are made of? Morgan did this by mating fruit flies differing in several traits (such as eye color, bristle character, wing shape, and so on) and determining how frequently the traits reappeared together in progeny. He discovered early on that some traits appeared more frequently together in progeny than could be accounted for by a completely free and independent reassortment of the kind Mendel had found. That is, the traits appeared to be *linked,* as though some genes were physically tied together. Because the chromosome seemed to be the place where genes were actually located, Morgan concluded that linkage between genes meant that they were *on the same chromosome.*

Morgan then crossed flies having different linked traits and found that different pairs of linked traits showed up in the offspring with different frequencies. By carefully working out the frequency of reappearances of various pairs of traits in progeny, he reached the conclusion that the farther apart trait-determining genes are on the chromosome, the more frequently they reappear in the progeny. This remarkable conclusion meant that genes must be arranged along the chromosome in a simple linear order. This basic genetic method that Morgan invented for determining the relative positions of genes in the chromosome became widely applicable in genetics and will be discussed more fully in Chapter V.

If genes are linked, that is, physically tied together on a chromosome, then we need to know why the several genes that Mendel had studied seemed to rearrange independently, as though there were no chromosomes. The genes he had studied seemed to have been *un*linked. This is one of science's great pieces of luck. The fact is, Mendel just happened to choose a group of traits to study that were, in fact, unlinked; their genes were on *different* chromosomes, or at least a long distance apart on single chromosomes, so they reassorted independently, unimpeded by linkage. If Mendel had chosen other traits, he would have obtained much more ambiguous results that would have been harder to interpret — by him and by posterity. He might then have remained in obscurity forever!

Morgan won a Nobel Prize in 1933 for "discoveries relating to the hereditary functions of the chromosomes." He worked in his laboratories right up until the end of his life in 1945. Without yet knowing what a gene was, he made it possible to envision it as a discrete part of a chromosome, arranged in space linearly with respect to its fellow genes.

Hermann Muller, a student of Morgan, was a brilliant, courageous, and far-seeing scientist. He worried about what made genes change, that is, what caused mutations. He became convinced that some physical force was at work, and that the force had an exquisitely sharp point, in that a mutation could change one gene and not touch a thousand others in the same cell. In casting about for a physical agent that could penetrate deep into the protective barriers of the cell and the nucleus, and have incisive action on a single gene, he struck upon x-rays. He then did an experiment in which he irradiated hundreds of fruit flies with x-rays, mated them with unirradiated fruit flies, and looked for mutations in the offspring. His harvest was rich indeed; every imaginable kind of mutation appeared in a large percentage of the offspring. There were flies with flat and dented and bulging eyes, in a variety of colors; flies with no eyes at all; flies with curly, ruffled, fine, coarse, forked, and no hair;

flies with no antennae; flies with broad wings and down-turned wings, outstretched wings, truncated wings, and almost no wings at all. There were big flies, little flies, active flies, and sluggish flies. There were short-lived and long-lived flies. It was as though he'd released the creatures from a Hieronymus Bosch painting or from a tiny Pandora's box.

How did these inherited changes compare with "normal" inherited changes, normal mutations? Apparently, only in number. The effect of x-rays was like the effect of natural mutation-causing agents, but more powerful. Muller also found that x-rays could in some cases reverse previously induced mutations, literally correct them. This seemed to mean that mutations were sometimes very small in extent.

Muller's great work had two major impacts: it shed light on the nature of mutations, and it provided genetics and biochemistry with an exceptionally valuable mutation-making tool for studying genes. Deliberately created and selected mutant organisms came to be among the most valuable objects for the study of genes. If there remained any doubt about the physical basis of genetic phenomena, this demonstration that radiation could permanently alter heredity was a giant step in removing it.

Muller won the Nobel Prize in 1946 for his work. He was widely recognized as one of the great intellects of biology. His vision was extraordinary. Way back in 1921, for example, he learned of the discovery of viruses called *bacteriophages* that attack bacteria. He imagined the system to be ideal for exploring the nature of the gene, and was quite specific in his suggestions as to what might be done experimentally. Twenty years later, the bacteriophage became the rallying system of the new biology.

It is surprising how much could be learned about genetics

without having any specific idea of what genes actually were or did. Mendel, Morgan, and Muller had convinced all who explored these matters that genes were the inherited units (materials, things, factors) that caused creatures to have certain specific physical traits and characteristics. But they had no inkling of what genes were or how genes determined these traits.

Genes had to be something, and that meant something *chemical,* for the whole body is chemical. Because genes governed complex functions, they had to be big and they had to be complex. Looking about in the body for gene candidates, scientists saw only two kinds of chemical structures that might be able to do what genes do. One was protein and the other was DNA. By 1944, DNA had been known for nearly 100 years, and a lot of good chemical research had been done on it. It had been discovered and described first by a German biochemist, Friedrich Miescher. It was known to be a surprisingly simple structure: some fairly simple chemicals put together like a string of beads. The business part of each unit is called a *base,* and there are only four bases:

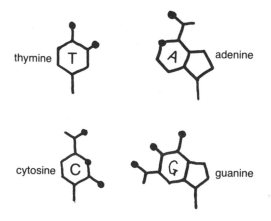

The angular ring shapes of the bases represent real rings of linked carbon and nitrogen atoms. The black dots identify a function we'll consider later. Each base appears in DNA attached to a sugar molecule (called *deoxyribose*) and a *phosphate* molecule:

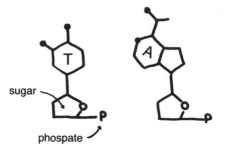

The sugars are rings of carbon and one oxygen, and the base is attached directly to the sugar. These fundamental base-sugar-phosphate units of the DNA chain are called *nucleotides,* a term we'll use frequently in this book. When these nucleotides are linked together, they make DNA:

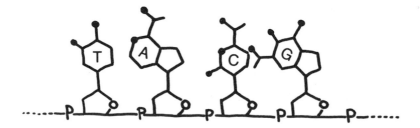

If you imagine the four nucleotides of DNA to be analogous to the 26 letters of the English language, then DNA is like a book from which the words, sentences, and paragraphs have been cut and pasted end to end to give a very long strip of letters. Four letters that can be arranged without restrictions on their order or the length of words or sentences will result in an infinite number of possible sequences. The implication is that DNA has meaning, just as the sentences in a book have meaning, but written in an almost limitless sequence of nucleotides.

DNA chains are stupefyingly long. The DNA of a single virus is 200,000 nucleotides in length; that of a bacterium, 2 million nucleotides long; that of one human cell, 1 billion nucleotides long. To go back to our book analogy, the bacterium's DNA would be equivalent to 20 average novels of 100,000 words; the human's, to 10,000 novels. Human DNA is broken up into 46 chromosomes in each of our 60 trillion body cells. If all the DNA of *one* human cell were laid out straight, it would be about a yard long. If all of our cells' DNA were laid end to end, it would reach to the sun and back many times! The only way such an enormously long thing can fit into the tiny nucleus of a cell is to be vanishingly thin. And DNA is just that.

If *big* were an important criterion, DNA would seem to be *big* enough, or at least long enough, to have a role in heredity. But was length enough, particularly when there were only four different kinds of units making up that length? One needed, really, to look at the other candidate — protein — to see what its qualifications were. The fact is that, before 1944, almost all scientists favored protein, not DNA, as the genetic material. And not without justification.

It was true that chromosomes resided exclusively in the cell's nucleus and that chromosomes were intimately involved in inheritance. Furthermore, all the cell's DNA was in the chromosomes. But lots of proteins were found in the nucleus, too. Not exclusively, like DNA, but they were there nonetheless. So they could be the genes. Proteins were also seductive for a more compelling reason. Proteins came in a variety of shapes and sizes. They were also chains, but chains of 20 different kinds of units (called *amino acids*) instead of only four. Proteins, then, were big enough and seemed to be complex enough to be candidates for the job of gene. By comparison, DNA seemed tiresomely monotonous and uniform, with only four units repeating themselves over and over. All molecules of DNA were pretty much alike, too. Could something like that carry all the information of inheritance? There was a tendency to relegate DNA to some subordinate role such as cell building material.

You see, nobody at that time caught on to the idea that the variety and diversity you needed to carry all the information of inheritance might be *hidden* in the molecule in some arrangement of the chain's links that is not obvious in a search for gross surface diversity. The reasoning that rejected DNA was like reasoning that would reject the "tape" we made by cutting out the printed lines of a book and splicing them end to end. The tape is long, monotonous, and boring. As a long strip of paper, it has little to offer someone searching for meaning and variety. But, of course, it is the *sequence of letters* that is loaded with meaning. That was the hidden message of DNA that had not yet occurred to biologists.

Think of it another way. Think of DNA as a single large book on your library shelf, say, the Bible. Outside, it's just a somber black book, but the sequence of its letters and words inside is full of meaning. Proteins, then, would be the rest of the books on your shelf, mostly smaller and varying in size, color, and shape. Of course, they would be full of letter sequences, too. It is understandable that scientists, who were not yet able to look too deeply inside any of these books, might see in the variety of secular volumes something analogous to genes.

Each idea and great venture has its time and place. After the fact, it is sometimes embarrassingly easy to say what should have been done, what the crucial experiment should have been. This is so in the case of identifying the substance that was the gene. In retrospect, all that was needed was to take some protein and some DNA from an organism, inject each into some genetically different form of the organism, and see whether the latter's inheritance was changed.

But of course things aren't that easy. And to my knowledge, nobody did such an experiment. One reason was that knowledge tends to be compartmentalized. Those who work in one area tend to stick to that area. Thus, the manipulators of molecules, the chemists, had almost no contact with geneticists, whose language, ways of thinking, and experimental methods and systems were different. The other reason was technical. The right test system simply hadn't been developed. The right test system — the model — is absolutely critical to the forward movement of science.

It fell to a scientist named Frederick Griffith, in 1928, to discover the model, the pneumococcus. The pneumococcus is the bacterium that causes lobar pneumonia. Before the days of antibiotics, lobar pneumonia (popularly called double pneumonia) was the major cause of death worldwide. If you managed to evade other ills, pneumonia could be counted on to finish you off in the end. Medical science was much concerned with the study of how immunity to the deadly pneumococcus was determined.

There are two main forms of these pneumococci, called *smooth* and *rough*. When you drop a few cells on nutrient jelly in a glass dish, they multiply, as is the way of bacteria, over and over until 10 million or so cells can be seen as a glistening, *smooth* colony about the size of the head of a pin. A colony is a *clone:* several million bacteria that have all come from one bacterium by repeated cell division. When you look at a sample colony under a powerful microscope, you can see that it is covered with a coat of fine velvet-like hairs. This coat is made of polysaccharides, which are strings or chains of sugar molecules. The coat is what gives each bacterial strain its unique identity, and the coat and associated properties are inherited from generation to generation. (Polysaccharides are important not only to bacteria. *Our* cells have polysaccharides on their surfaces, too, and they play a vital role in cell identity.)

Rough pneumococci, in contrast to smooth pneumococci, have lost the ability to make the characteristic polysaccharide coat. Under the microscope, the colonies of rough pneumococci are ragged looking and the cells are devoid of a coat. The inability to make a coat is inherited just as the ability to make a coat is. Now the pneumococcus in a smooth coat is the kind to avoid. The coat confers virulence, the ability to cause disease and death in animals and humans. The rough bacteria are quite benign: no coat, no disease. The pneumococcus' danger then lies in its coat. Inherited by each new generation, that coat gives the tiny organism deadly dominion over man and mouse.

Griffith, an intensely dedicated scientist, was determined to unravel the mysteries of pneumonia. An excruciatingly shy man, he worked in the Ministry of Health in London and devoted his life to the study of pneumococcus. In 1928, he was trying to discover why smooth pneumococci recovered from patients with pneumonia differed according to the part of England in which the patient lived. In particular, he was trying to see whether factors inside a human or an animal might cause pneumococci to change from one smooth type to another smooth type.

As part of his search, Griffith set up an experiment. He injected simultaneously into mice a small number of benign rough cells and a larger number of dead smooth cells. These smooth cells had all been killed by heat and could do no harm to mice, because dead bacteria, of course, don't cause an infection. Griffith was surprised to see that the mice got deathly sick! And in their blood he found *living smooth* cells. He had injected a few living rough cells and now had a rapidly multiplying population of *smooth* cells growing in the mice!

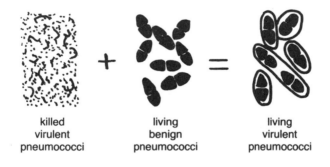

killed	living	living
virulent	benign	virulent
pneumococci	pneumococci	pneumococci

Clearly, unless we want to believe that the dead smooth cells were reincarnated, we must assume that the only living cells present, the rough ones, had been turned into *smooth cells* and had then multiplied, producing many cells all inheriting the smoothness trait. To us now, it is obvious that this transformation from rough to smooth cells was a genetic, hereditary event. It seemed that inside the mouse, stuff from killed smooth cells was changing the inheritance

of living rough cells. A truly astonishing discovery! But nobody, including Griffith, grasped its significance. In fact, he
died in the London blitz 13 years later while working in his
lab, and as far as I know, was unaware of the explanation for
his monumental finding. Bacteria at the time were thought
to be too small and primitive to have genes.

Griffith comes across the years as a sad, lonely figure, but
he obviously took much pleasure in his research. He explained his result by envisioning the smooth cells as providing a soup of polysaccharide in which the rough cells could
find the material to build capsules. His explanation would
have done damage to the generally held (and basically correct) idea that smooth and rough traits were unique and
unchanging and were passed on to progeny. The fact that
he had changed the inheritance of living organisms by exposing them to genes from dead organisms didn't dawn on
him. Instead of seeing the phenomenon as a curtain-raiser,
Griffith and most of his contemporaries saw it as dismay-
ingly disruptive by making things inexplicably more complicated.

But scientists followed Griffith's work with interest. He
was well established in his field and widely respected for
his careful work. After he published his transformation experiments in 1928, the work was quickly confirmed in Germany, China, and at the Rockefeller Institute in New York.
And then two important additional refinements were made,
both in New York. Around 1930, Martin Dawson and Rich-
ard Sia, at Columbia University, got around the need for
using mice. They succeeded in transforming rough cells to
smooth cells by exposing rough cells to heat-killed smooth
cells *in the test tube*. Mice, it turned out, hadn't really contributed anything to the process; they had simply been a

place where the bacteria could grow. They'd been used because they happened to be the test system being used at the time to imitate pneumonia in humans.

The second refinement was made by Lionel Alloway. In 1932, he took the stuff from the heat-killed smooth pneumococci (broken cells and the miscellany released from inside shattered surrounding membranes) and forced it through a fine filter. This held back all the large fragments, and what came through was a clear fluid. The clear fluid successfully transformed rough cells to smooth cells! The inheritance-altering chemical was, then, in the clear fluid where one might hope to identify it.

The stage was set.

CHAPTER TWO

The Chemical Nature of Genes and the Key to Their Action Are Revealed

Oswald Avery proves that genes are made of DNA. George Beadle and Edward Tatum show that genes govern the making of enzymes. Early studies of Archibald Garrod are brought to light that show mutations in human genes cause loss of enzymes.

IN SCIENCE, ideas are a dime a dozen. The really important ideas, which are rare and hard to come by, are the ones that can be tested, that point clearly to an experiment that will give an unequivocal answer to a well-formulated question. The Griffith discovery now awaiting exploitation was the first stirring of the winds of enlightenment in molecular genetics. It offered an almost ideal experimental system: a simple organism whose inheritance could be changed by stuff taken from another organism. The phenomenon had been named *transformation* by Griffith. It offered the opportunity both to manipulate inheritance and to find out what it was that was causing the heritable changes.

Genetic knowledge is ultimately applied in medicine and

agriculture, in the treatment of disease and the improvement of agricultural products. But humans are too complex and inaccessible to be useful experimental subjects, except in rare instances. Plants and animals, though better, are still complex, variable, and inconvenient. So simple models of life, usually found to operate as the more complex systems do, are sought to facilitate exploration of mechanisms. Those simple pneumococci were particularly enticing to the curiosity of Oswald Avery, who saw them as ideal models for genetic experimentation. He was uniquely situated to take the baton from Griffith.

Avery was a physician by training but by professional interest, a microbiologist (a specialist in small forms of life like bacteria and viruses). He worked at the Rockefeller Institute in New York City. He had devoted his career to the pneumococcus, particularly to understanding the body's immune defenses against the organism. Part of the body's way of defending itself against disease is to deploy antibodies. Antibodies are special proteins that have the unique ability to bind firmly to parts of bacteria, thereby making the bacteria more susceptible to destruction by other means the body can muster for the occasion. One of Avery's important discoveries in the 1920s was that the body's immune defense system recognized the polysaccharide coats of the bacteria and made specific antibodies to them. Until then, it had been thought that antibodies were made only against protein molecules.

At first, Avery was skeptical about Griffith's experiment because the results, as interpreted by Griffith, did not fit with current knowledge. But the rapid confirmations of Griffith's work and the technical improvements made by Avery's own colleagues soon convinced him that an exciting system was at hand. All of these events helped to launch Avery on a ten-year journey, from 1933 to 1943, that finally revealed to him the secret of bacterial transformation. For seven years of that time he was accompanied on the search by two valuable associates, Colin MacLeod and Maclyn McCarty, whose names would appear beside Avery's on the 1944 publication that described their successful work in the *Journal of Experimental Medicine*.

Avery's task was tough. The fluid obtained from killed smooth pneumococci contained thousands of different kinds of protein molecules, all sorts of other substances, DNA, and RNA (about which we'll hear more later), that is, all of the stuff essential to the life of the pneumococcal cells. It was like a rich soup. From that brew he would have to select one class of molecules and show that they alone were responsible for transformation. The testing involved exposing rough pneumococci to the various molecules he isolated to see whether they would be turned into smooth pneumococci. This was proof by elimination: proving that something was important by showing that a lot of other things were not. If he guessed that one particular kind of molecule — say DNA — was the right one to focus on, no one would believe it was responsible for transformation unless it was absolutely *pure*. If it was contaminated with anything else, the proof would not be convincing; the "contaminant" might be the substance causing transformation.

Avery and his colleagues were helped by the methods used by Alloway to prepare the transforming fluid. They first killed the virulent cells by heating them. This also inactivated some of the enzymes inside the cells that could destroy the material responsible for transformation. (Enzymes, the machinery that carries out all cell functions, are special proteins that act as chemical catalysts, pushing chemical reactions forward. Enzymes may be removed from cells, purified, and used as reagents, identical to chemical catalysts. Even the simpler cells — for example, pneumococci — contain thousands of enzymes.) Then they added chloroform to the fluid so as to remove most of the *protein* from the extract. The carbohydrate (polysaccharide) material of the coat of the bacteria, associated with the property of virulence, was removed by an enzyme that broke it down to its component small sugars. What remained after these and other manipulations was repeatedly purified by precipitation with alcohol and redissolving in water. They tested bits of the fluid at each step along this course of purification and found that the steadily less complex fluid continued to cause transformation. At the end of the long trail was a fluid from which almost all of the cell's material had been removed. It looked like water and could change inheritance. What was in it?

It wasn't *carbohydrate*, because that had all been destroyed. It was fairly certain that it wasn't *protein* either, because essentially all of that had been removed by the chloroform treatment. But to make sure, Avery also added pure enzymes that destroy proteins; these had no effect on the transforming ability of the fluid. Very sensitive specific immunologic tests to detect carbohydrate or protein were also negative. On the other hand, all the tests made of the transforming material gave the same answers as were given by samples of known DNA that Avery obtained from DNA specialists — samples that were as pure as any available. Finally, treatment of the transforming fluid with an enzyme that breaks DNA apart, completely eliminated the transforming activity. The scientists were at last convinced that the genetic material of the pneumococcus responsible for its transforming ability was DNA.

Starting with a mixture of stuff from cells, doggedly eliminating one thing after another, and finally concluding that DNA must be the transforming agent because nothing else was may not seem the stuff of adventure. Perhaps never was such a momentous discovery made in such an undramatic way. And momentous it was, as our story will show. To be the first to hold a test tube full of pure genes and to be able to envision its power to bring new knowledge must have moved Avery profoundly.

Most of the scientists whose work we'll discuss in this book won Nobel Prizes for their achievements. Avery, who led the grand procession, was left out. Part of the reason was that he was older than his colleagues when he completed his work; he was not one of the "in" set. Most of his career had been lived out before the new biology drew up its battle lines in the 1940s. And Avery had some of Griffith's qualities: he was the textbook model of the old-fashioned scientist letting the facts speak for themselves without rhetoric. Then, too, scientists were reluctant to accept DNA as the genetic material. A majority favored protein for the reasons discussed in Chapter I. Avery's proof could not absolutely exclude the presence of small amounts of protein in the fluid. Skepticism, generally a healthy trait in scientists, bordered on the absurd in this case.

The protein bias limped past its time by virtue of a loophole in Avery's experiments: protein couldn't unequivocally be ruled out. But the skeptics were not members of a scientific old guard, or establishment. They were the new elite who would launch molecular biology; the brilliant, cocksure, and arrogant "phage group," the avant-garde of the new biology!

Avery would probably have got the Prize if he'd lived a few years longer, but he died in 1955. In his last ten years he was as aware of the significance of his discovery as he was reluctant to dispute with his critics. He knew he'd made a great breakthrough at the end of his career and he took satisfaction in the solid contributions of his earlier work on the pneumococcus. The doubters disappointed him, but there were other scientists who immediately felt the rightness of Avery's experiments and took his lead into new realms of exploration. We shall learn about them soon.

✿

Eight years after Avery's 1944 publication, an experiment was done by a charter member of the phage group that was widely and quickly accepted by scientists as definitive and conclusive evidence for the role of DNA as genetic arbiter. The system used by Alfred Hershey and his associate Martha Chase in their laboratory at Cold Spring Harbor, Long Island, was a particularly exciting one for genetic exploration. Bacteria are subject to infection by viruses, called bacteriophages, or *phages* for short. Phages are made of protein and DNA, and we know now that they act like tiny hypodermic syringes, attaching themselves to the outer surface of the bacteria and then injecting their contents — mostly DNA — into them.

Amazing things then happen inside the bacteria. First, while metabolism continues, all machinery for bacterial growth stops; "silence" reigns. A stealthy reprogramming of the bacterial machinery proceeds and is soon evidenced when new phage parts start to appear. About 100 copies of phage DNA are produced, as well as 100 syringe heads and needle parts made of protein. The DNA then finds its way into the protein heads, and 100 completed viruses now inhabit the interior of the bacterial cell. The cell then forms an enzyme that dissolves its own outer membrane and bursts, releasing the viruses to attack other bacteria. In this way, infection rapidly spreads to kill all the bacteria in the culture.

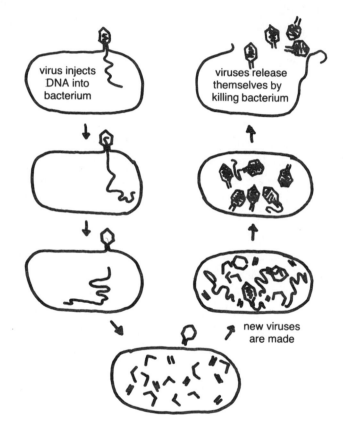

It took a new breed of scientists, the phage group, founders of molecular biology, to see in the phage system the potential for shedding new light on the mechanism of life. Both the phages and the bacteria are simple organisms, yet their life-and-death struggle represents potent inter-plays of molecular forces. The burning question was: How was the phage able to sabotage the bacterium's life proc-esses in the furtherance of its own interests? At the time the expectation was that the whole virus entered the cell, in-fecting it just as bacteria enter us and infect us.

What Al Hershey and Martha Chase did was to allow phages to infect bacteria. The suspension of phages and bacteria was then vigorously shaken to remove adherent phages from the bacterial surfaces, and the amounts of phage protein and DNA associated with the bacteria were measured. Most of the DNA was with the bacteria, whereas most of the protein had been washed away. Hershey and Chase concluded that DNA was the material that entered the bacterium from the phage and was responsible for the subsequent dramatic events inside the bacterial cell. That conclusion was hailed by the new biologists.

But the Hershey-Chase experiment wasn't as conclusive as it sounds when the actual quantities were closely scru-tinized. Although 80 percent of the phages' DNA did in-deed stay with the bacteria after the phages were removed, 20 percent did not! And although 80 percent of the protein was removed with the phages, 20 percent was not! Twenty percent of anything is a lot. The experimenters attributed the discrepancies to technical difficulties and put their money on what *most* of the DNA and protein were doing. They proved to be right in the end. The readiness with which Hershey and his phage group colleagues accepted the results of the experiment can in the end be attributed to Avery (and Griffith). For in the eight years since Avery's 1944 publication, his critics had shed their prejudices and had come to look open-mindedly at his work.

The Avery-Hershey experience points up a couple of interesting generalizations about science. One is that new ideas can languish until the right moment for acceptance is at hand. The other is that conclusions are more acceptable when derived from familiar systems. The leaders of the new forces of biology were, because of their backgrounds, more ready to accept a weak experiment on phages than a solid experiment in cellular biochemistry.

❖

Whatever explanation would eventually emerge for the control of heredity, it would have to find a central role for this long, thin chain of four repeating nucleotide units.

With the identification of DNA as the actual chemical material that genes are made of, the door seemed ready to be opened on the vast unknown area of gene action, the science of genetics. But the genetics of Mendel, Morgan, and Muller lacked precision. Theirs was a genetics of unknown entities governing ill-defined traits. A chemical key was missing: what was the gene actually *doing?* What underlying chemical mechanism was responding to the instructions written in the gene?

In one of those dramatic coincidences of science, only two years before the publication of Avery's work, a geneticist and a biochemist working together were completing experiments that would answer this question of gene action. Thus, the chemical identity of the gene and the chemical nature of the gene's action were discovered at almost the same time.

The geneticist was George Beadle; the biochemist, Edward Tatum. Together they opened vistas that would have been closed to either separately. The extraordinary discoveries of these two men, which would later bring them Nobel Prizes, were made at the California Institute of Technology. Beadle had been brought up on a farm in Wahoo, Nebraska, and had been persuaded by a high school teacher to enter science. He got interested in genetics at Cal Tech, where Morgan's school of genetics had earlier been transplanted from Columbia and was being continued in the traditions Morgan had established.

Following up on experiments begun by others, Beadle started his work trying to discover how tissues from an organism of one particular genetic make-up could affect the expression of genes in an organism of another genetic make-up. For example, he discovered that if he transplanted tissue (i.e., cells) from normal fruit fly larvae into larvae having a mutant defect in eye color, the flies that developed all had normal eye color. Beadle concluded correctly that the normal cells secreted some chemical substance that was required for normal eye color and that this substance was lacking in the mutant. He eventually isolated and identified the particular chemical involved.

Beadle saw then that a gene was responsible for *producing a chemical* that in turn produced the trait the gene was identified with. The importance of this insight, stated so easily, cannot be overemphasized. Even the simplest trait or characteristic of an organism is a composite expression of many things and occurrences, all of a chemical nature. Beadle opened the way to seeing the gene as a factor controlling *one* chemical thing or event. This narrowing of focus was the very essence of the approach to the truth about gene action.

Beadle was now joined by Edward Tatum, a biochemist. They theorized that, because no chemical substance can be made in the body without the aid of enzymes, *genes must control enzymes.* (Remember that enzymes, which are protein molecules, control all cellular functions.) They imagined that a gene caused an enzyme to be made, although there was little evidence to support this idea at the time. The idea's logical extension was that *all* traits were orchestrated by enzymes. Traits produced by many proteins and enzymes were, consequently, the product of many genes. It was an enormously intriguing, powerful, simplifying stimulus to researchers in the field. And it had the added merit of turning out to be right!

Beadle and Tatum changed to a less complicated and far more useful model system to test their ideas, the common bread mold *Neurospora.* They could work with traits that were simpler than eye color or wing shape, for example, the loss or gain of the organism's ability to grow on some nutrient. That was a more chemical, less visual, narrower trait more likely to be under the control of a single gene. Borrowing from Muller's discoveries, they irradiated bread molds with x-rays to produce mutants. Normal molds are versatile in the capacity to make most of their own constituents. Mutant molds were readily identified because they had lost the ability to make some essential constituent. The result was that new types of organisms appeared which failed to grow in a nutrient solution that lacked many essential constituents. By adding nutrients back one at a time until the new organisms started to grow, Beadle and Tatum could discover what compound the organisms needed.

Let's look at an example of the kind of experiment they did. *Neurospora* seeds were x-rayed. They were then grown in a medium of salts, sugar, and one vitamin on which normal *Neurospora* can grow but mutants cannot. If a mold couldn't grow on this, they assumed it had been damaged in an enzyme essential to converting the simple substances available in the medium to the substances the organism needed and could normally make for itself. In one such experiment they found three mutant molds that could be made to grow again by adding to the medium the amino acid arginine (mutant 1); arginine *or* another amino acid, citrulline (mutant 2); and arginine *or* citrulline *or* a third amino acid, ornithine (mutant 3).

These experimental findings led them to postulate that the normal manner for making the cell's needed arginine was: unknown substance $\xrightarrow{3}$ ornithine $\xrightarrow{2}$ citrulline $\xrightarrow{1}$ arginine, and that mutant 1 had lost the ability to accomplish reaction 1; mutant 2 was similarly blocked in reaction 2; and mutant 3 failed to carry out reaction 3. Because each of these reactions is accomplished by a very specific enzyme, Beadle and Tatum concluded that gene 1 somehow controlled the integrity of enzyme 1; gene 2 controlled enzyme 2; and gene 3 controlled enzyme 3.

Beadle and Tatum were able to induce all sorts of new nutritional requirements in the mold by damaging different genes. Because there are many hundreds of compounds a versatile organism can make, and because each is made by a specific enzyme, there are an enormous number of nutritional defects that can be generated by mutation. Beadle and Tatum's experiments led them to enunciate the famous *one gene–one enzyme hypothesis:* each gene is responsible for the production of a single enzyme. This became the frame within which the picture of genetics took shape in the coming years.

✻

Classical genetics ended and the new biology — molecular genetics and molecular biology — began with the demonstration that genes make proteins and that genes are made of DNA, discoveries published between 1941 and 1944. Geneticists were going to change their emphasis, looking less at the next generation, and more at what genes did in the present in the very cells in which they resided. The manner in which genes control the expression of a cell's potential became the experimental focus.

Most biological research is, in the last analysis, directed toward understanding ourselves. Bacteria, bread molds, fruit flies, and peas are simply convenient, if surprising, models of ourselves. They are used, of course, because it is seldom efficient or convenient to study ourselves. We're too complicated, for one thing, and we don't look kindly on human experimentation. There are, however, a few exceptions.

One exception was a scientist and physician who worked in England, who "scooped" Beadle and Tatum by about 30 years. He was Sir Archibald Garrod, a brilliant clinician and perceptive observer. His observations on his patients led him to exactly the same conclusions reached by Beadle and Tatum. He found that a number of unusual hereditary diseases in his patients were the result of specific defects in the ability of their bodies to carry out simple normal biochemical functions. Because he knew these functions were mediated by enzymes, he concluded that his patients had inherited defective, that is, mutated, genes that failed to produce the corresponding enzymes. The enzyme defect, resulting in inability to make some chemical needed by the body, was the cause of the illness. Garrod's work has since been copiously confirmed in the study of more than 200 human genetic diseases. Beadle only became aware of Garrod's achievements late in the course of his own studies. He then took it upon himself to become Garrod's reviver, doing much to make the scientific community aware of his remarkable foresight.

Garrod's premature vision of the truths that became the central focus of biology more than half a century later is reminiscent of Mendel's. Scientists' receptivity to germinal ideas depends on intellectual fashion, on technical readiness, and perhaps on other variables we know little about. Knowledge can lie buried until others are ready to dig it up. The submersion of Garrod's work contrasts sharply with the immediate success of another germinal study of genetic disease of humans — sickle cell anemia — which came when the time was ripe. Although lowly peas and fruit flies and bread molds and bacteria have built the bulk of genetics, the participation of human subjects in the drama has provided the required high level of confidence that truths revealed by simpler forms of life are applicable to the more complex forms.

CHAPTER THREE

Bacteria as Models

The characteristics of bacteria that make them valuable experimental subjects. The place of bacteria in evolution: Lamarckian inheritance of acquired traits versus mutation and selection. Salvador Luria brings bacteria into the Darwinian fold.

IN THE EXPLOSION of knowledge of life processes that followed Griffith's and Avery's slow-fused revelations, bacteria came to play a major role as experimental subjects. They were asked most of the big questions and they answered well. That is to say, bacteria served as an excellent *model system* because they were simple and easy to work with, grew rapidly, and yielded answers to questions that could be applied to all living creatures.

It would be reasonable to assume that Avery's success with the pneumococcus inspired the growth of bacterial experimentation, but it didn't. Bacteria became the "in" experimental system of molecular biology in the mid-1940s owing in large part to the experimental ingenuity and promotional skill of a small coterie of scientists called the phage group, who were seen briefly, not in their best light, in Chapter II. With their primary background in physics, these scientists entered biology with a strong conviction that life principles could be understood by applying physics and chemistry to the *simplest organisms*, namely, bacteria and viruses. They proved to be resoundingly right, and molecular biology came to be defined as the combined use of physics and chemistry and genetics in the study of life processes. The phage group was not a physically cohesive group, but rather was a handful of highly articulate scientists in this country and Europe who kept in close touch by mail and meetings. They shared ideas, intuitions, prejudices, and expectations and inspired a few exceptionally able students. The group's leader was a physicist named Max Delbrück.

In 1943, a dramatic experiment was generated by the phage group's intellectual exchanges. The experiment was conceived in an inspired flash and its result was clean and definitive. Before 1943, microbiologists had seen variation and adaptation in bacteria as fundamentally different in mechanism from the same phenomena in higher plants and animals. After the 1943 experiment, bacteria were seen as one with all the rest of living creatures. Genetics and even evolution could be studied in bacteria and the results considered applicable to humans.

Let's set aside for the moment the question of whether bacteria and animals obey the same laws of genetics and evolution and consider other factors that make bacteria better models than mice or rats, which, after all, are much more like humans.

First, bacterial cell populations are genetically *pure* strains. By pure I mean that every cell is identical. If you start with a single cell, all cells in the population resulting from that cell's division will be identical. You will recall from Chapter I that the process of growing a mass of identical cells from a single progenitor is called *cloning*. So you have a population of cells all the same, all doing the same thing, sometimes even at the same time. That uniformity is important to the scientist, who always seeks to work with as pure a system as possible. It's much harder to get genetically pure mice or rats, and when you do, they're much more complicated than bacteria.

Second, the *numbers* of cells you work with are enormous. It's easy to obtain many millions of cells from a few, and they take little room. The importance of numbers lies in the obvious practical need to have enough cells to work with, but also in the need to search for and detect *rare events* that occur in only a few cells among millions. This ability to *select* rare organisms is critical to the success of microbial genetics.

Third, bacteria grow *fast*. Under the most favorable conditions, bacteria make a new generation every 20 minutes! This means you can produce hundreds of generations in the lab within a few days.

When I took microbiology in medical school, I was appalled to learn of the implications of bacterial growth rate, of the enormous mass of cells that can, theoretically, be produced in a short time. The mathematics are simple: if you start with 1 cell and it divides in 30 minutes, you have 2 cells in half an hour, 4 in an hour, 16 in two hours, 64 in four hours, 16,384 in eight hours, and so on. In a very short time you'd have billions and trillions of bacteria inheriting the earth and swamping out all other forms of life. This type of growth, by which the numbers regularly double, is called *exponential* growth. Fortunately, bacteria can't grow indefinitely because they run out of food or they poison themselves with their own waste products. If they didn't, starting with one cell on Monday, you would have a layer of bacteria a mile deep covering the earth by Wednesday!

Bacteria not only grow fast, they do everything fast. They make things and stop making things fast, they adapt fast, and they express a change in their genes fast. You don't have to wait around for things to happen.

These properties of bacteria might make them fun to experiment with, but what would give a scientist reason to value them as *models,* that is, as analogues of higher organisms? The scientist would want to ask whether bacteria follow the same laws of genetics and evolution that we do. Their usefulness as models hinges on whether they are like us in the fundamental rules governing their existence.

Charles Darwin, of course, led the way to our understanding of variation in populations of plants and animals. By the 1940s, biological variation was understood in genetic terms, even though its molecular basis was not known. The primary instrument of change was the mutation, a change in a gene. In addition, sexual reshuffling of genes greatly amplified variation. Altered genes, or new gene combinations, inherited by subsequent generations, caused organisms to manifest new or different traits. Underlying these altered traits was an altered or damaged enzyme, as Beadle and Tatum had shown. These variations in an organism's traits were the instrument on which the environment played. If a variation improved an organism's ability to survive in a given environment, and therefore to produce offspring, the organism flourished, having been favorably *selected* by the environment.

For example, a random gene change that improved an antelope's running speed would give that member of the antelope population a better chance of surviving in an environment harboring fleet-footed predators. The underlying cause of improved running ability may be found in muscle function, skeletal action, use of energy, and so on, any of which will be mediated by proteins and enzymes under the control of genes.

Oddly, before 1943, scientists saw bacteria as exceptions to these rules. They thought that variability in bacteria had some nongenetic explanation. We know bacteria are the most common form of life on our planet, if our measure is total mass of living material. That they should have been considered exceptions to the laws known to govern all other forms of life seems strange to us now. Scientists working with bacteria before the 1940s were narrowly focused, seeing bacteria as their special domain. Bacterial variation and adaptation seemed to them to occur too rapidly to be explained by Darwinian evolution. They noted, too, that bacteria didn't have nuclei, didn't appear to have chromosomes, and made more of themselves by splitting in two without exchange of genes. They simply were not seen as participants in the evolutionary drama. It seems that, focusing on bacteria for their own sake, the early microbiologists failed to see their subjects as part of the total biological continuum. The new breed of biologist, exemplified by the phage group, was out to unify biology in genetic and molecular terms and so tended to see bacteria as a piece of the whole, and as potentially valuable tools (models).

Furthermore, an increasing number of biochemical studies were showing that bacteria were composed of all the same chemicals that other organisms were, such as amino acids, nucleotides, vitamins, and so on. This suggested they shared a common evolutionary background with higher organisms. As we learned more about bacterial internal mechanisms, we saw they were similar to our own. Bacteria that could subsist on a much simpler diet than we needed turned out to be able to make many of the necessary substances themselves. They had more complex machinery than we had as far as nutritional needs go. So simplicity of nutritional needs was not a reflection of simplicity of internal machinery, but rather a reflection of evolutionary gain in the ability to make what they needed.

An outstanding feature of bacteria that impressed all who studied them in the early days was their adaptability. It seemed that whatever circumstances they found themselves in, no matter how adverse, they were very quickly able to adjust, survive, and thrive. Thrive, of course, meant multiply. Let's look at an example. If bacteria were growing vigorously and you added a drug that killed them, you'd think that would be the end of them. But it wasn't. Although almost all of the cells were killed in a short time, new bacteria arose that grew vigorously *in the presence of the drug*. They were drug resistant! (This kind of change to resistance doesn't happen very often in a population, but because there are millions and millions of bacteria to work with, we can always find the rare resistant bug simply by growing large numbers of bacteria in the presence of the drug. The few resistant ones will survive and multiply, whereas the rest of the cells will be killed.)

Bacteria also respond appropriately to kindlier treatment. Add to their fluid something they like to eat and they respond with a swelling population. Eating a simple diet of one sugar and a few salts, bacteria can manufacture all the complex building materials they need to make more of themselves. If you give them the compounds they usually make, they rapidly stop making them and use the gift. When cultivated in the laboratory, some pathogenic (disease-causing) bacteria are slow and finicky in their growth at first, but then begin to grow faster. As they adapt to laboratory conditions, their ability to produce disease in animals is weakened. There seems to be a trade-off. Why waste lethal skills when no victims are around? Best to adjust to lab living.

Bacteria, then, present a picture of rapid and dramatic adjustment to a great variety of environmental conditions. The traditional microbiologists did not relate that kind of behavior to genetic and evolutionary phenomena in higher organisms, which seemed generally much slower and more conservative. The rapid adaptability of bacteria compared to the stability and resistance to change of higher organisms didn't seem explainable by the same mechanisms.

*

Salvador Luria came to the United States in 1940 as a refugee from Italy, where he had been a physician. Unlike his close associate for many years, Max Delbrück, he was not a physicist. Delbrück, who had fled from Germany in 1937, and Luria became the founders and leaders of the phage group. They were joined in 1943 by Al Hershey, an American. They were soon to recruit an increasing number of brilliant, colorful men who set the pattern and pace of biology for the next two decades. Delbrück, Luria, and Hershey won Nobel Prizes in Medicine in 1969 for their work in bacteriophage biology.

Luria ruminated on bacterial adaptation. Is the bacterium's ready adaptability something that the environment induces? Are the bacteria modified directly by the chemistry of their surroundings? Or do many changes occur randomly in the bacterium's genes, with those allowing certain mutant bacteria to function better in a given environment being selected by that environment?

Luria knew that inheritance in higher organisms is determined by the interplay of chance changes in genes caused by mutation, and selection for or against the change by the environment. The more varied the genes in a species, the more likely it was that one variant of a trait would provide its carrier with the means to survive adversity.

Luria was deeply skeptical that bacteria were deviating from the Darwinian path to follow their own rules of inheritance and evolution. The idea that bacteria were somehow directly modified by their environment was the old discredited Lamarckian theory of inheritance of acquired characteristics. It would subscribe to the view that basketball players' children are taller than others because their parents were stretched by the game! Lysenko, the great charlatan of genetics, climbed out on that limb, taking Stalin and later Khrushchev with him, and succeeded in setting Soviet science and agriculture back 30 years.

Luria was out to bury Lamarck once and for all. He knew that phages, made of a few genes and proteins, readily attack susceptible bacteria, multiply inside them, and, in escaping, destroy the bacteria. Sometimes bacteria become *resistant* to phages; they simply are no longer bothered by the virus. This is a dramatic adaptation that provided an opportunity to do a critical experiment.

Luria saw that the key was to find out whether the change to phage resistance was something the phages did to the bacteria, or was the result of random, unpredictable changes, followed by selection. This second (Darwinian) explanation envisioned bacteria undergoing mutations constantly but infrequently, such that an occasional bacterium would be altered to phage resistance. Then, when phages appeared in the surroundings, the resistant cells could multiply as the rest of the cells died. Had no phages been around, the rare resistant bacterial mutants would never have been detected.

The problem can be simply stated in terms of experimental observation. Normally, when drug-sensitive bacteria are spread on a nutrient dish containing the drug, they are all killed. But occasionally, a colony (the progeny of a single cell) grows that is clearly drug resistant. Did the drug induce a change in a rare cell? Or did it just select an already changed cell?

How could Luria experimentally prove the Darwinian mechanism? This brings us to his 1943 experiment. Luria says the idea came to him while attending a University of Indiana faculty dance held at the Bloomington Country Club on a Saturday night. He was watching some of his colleagues play slot machines and was mulling over the fluctuation of the returns. It suddenly hit him that he could do a critical experiment in which the fluctuation of the results would be the key to his problem.

Luria set to work in his lab the next day. He put an equal amount of normal phage-sensitive bacteria into each of two 100-ml (cup-sized) flasks containing plenty of food. The contents of the second flask were immediately divided up evenly into 100 small 1-ml (thimble-sized) tubes. All cells were allowed to grow overnight. If a mutation occurred early during growth in one of the small tubes, its progeny would multiply many times during the night. In contrast, a change occurring late in another small tube would have only a short time to accumulate. Similar changes occurring in the big flask would not be so isolated, and so would be mixed into the whole population.

The next morning Luria set out 200 dishes, every one covered with a smooth layer of phages. On the first 100 phage dishes, he poured 100 small samples from the large overnight growth flask. On the second 100 dishes, he poured the contents of the 100 tubes of bacteria grown separately. All the bacteria grown up overnight, half in one big flask and half separated into 100 small samples, were now mixed

bacteria
immediately transferred
to 100 small tubes

bacteria grown
many generations,
then transferred to
100 dishes with phages

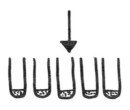

and grown many
generations. Then
transferred to dishes
with phages

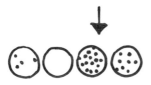

with phages. Now he waited for phage-resistant bacteria to grow. Wherever a single resistant bacterium landed on the dish, a colony would form. A colony, as we learned in Chapter I, consists of several million cells that all grew from one cell. So Luria could see how the conditions of prior growth in the absence of phages influenced the numbers of colonies that appeared in the presence of phages. In those numbers, he would find the answers to his question.

As Luria counted the colonies of resistant bacteria, he saw he was marking up another clear victory for Darwin. On the dishes from the 100 samples of bacteria grown separately, the numbers of phage-resistant colonies fluctuated markedly; many dishes had no colonies, some had a few, and others had many. This was the result expected if random, spontaneous mutations to resistance had occurred in different tubes at different times during the night. Early events, which would allow the most time for growth, would lead to many colonies; late events would lead to few. By contrast, the control part of the experiment showed only a small fluctuation in the number of resistant colonies on the 100 dishes from the single large flask of bacteria grown overnight. The differences from one dish to another simply conformed to the statistical expectations for samples from the population in the large flask.

Luria solved, literally overnight, the controversy over the origin of bacterial variation. He brought bacteria into the genetic and evolutionary fold with their more complex brothers and sisters, and he provided molecular biology with an enormously valuable new experimental tool — the bacterium — which would henceforth be the most widely used organism in the search for the details of gene action.

Luria's and Delbrück's experiment (Luria was an asso-

ciate of Delbrück and they published the work together),
clever and apt as it was, seemed a bit anticlimactic. The
prior conflict over the place of bacteria in nature and the
ingenuity of Luria's solution have elements of high drama.
But the outcome of the experiment is surprisingly simple.
We want to say we knew the outcome anyway. It is as
though St. George had laid a trap for the dragon and the
creature had fallen in while the hero was home eating
supper.

Beadle and Tatum, Avery, Luria. Within three years the
gene-protein relationship, the chemical material of the gene,
and the best possible experimental model for carrying on
the work were all revealed. No wonder the period of the
early 1940s is seen as the birthday of a new biology.

CHAPTER FOUR

Mutation and Human Disease

Mutations are changes in genes. Altered genes produce altered proteins. Linus Pauling links gene mutation to change in the protein hemoglobin in sickle cell anemia. Vernon Ingram pinpoints the hemoglobin change.

THE FINAL TRANSFORMATION of genetics to molecular genetics came with the first demonstration that a mutation in a gene caused a *specific identifiable chemical change in a protein.* Such a change in protein would destroy or seriously damage the protein's function, thereby accounting for the altered characteristic of the organism. Indeed, the geneticist was moving from a preoccupation with the functional effects — the outward manifestations — of mutations to the structural changes that mutations produced in protein molecules, a more refined and useful level of analysis.

The mutation-induced change in a protein was seen to *underlie* the more obvious change in function and therefore was closer to the center of events. It came about that mutant organisms were extremely valuable tools in analyzing cellular function generally, including protein structure and function. It was analysis of function by damage: wound a part of the cell, see what functional deficiency this causes, and then seek out the protein causing the deficiency. It was a bit like trying to learn what each part of your car's engine does by blindly smashing one part at a time, observing the resulting symptoms, and then finding and replacing the damaged part. A crude method for your car, perhaps, but one that has yielded much of the extensive knowledge we now have about cellular function.

Science was now focusing its curiosity on the relationship between the gene and the protein it produced. Altered traits or functions were now seen as secondary to changes in proteins. Mutations could alter proteins and such alterations could be clearly defined in terms of chemistry and physics. This produced valuable information on both how proteins work and how genes control proteins. As is frequently the case in science, the emergence of an explanation is accompanied by refinement of methodology that permits asking sharper questions. Thus in the new biology the mutant organism became a powerful tool for further analysis of living function.

Before we move further into the uses of genetics, our progress will be aided by some information about the nature of mutations and the structure of protein molecules. Mutations are simply changes in DNA. They occur spontaneously through inherent limits in DNA's defenses and in the fidelity of DNA replication. Their frequency can be increased by natural radiations like ultraviolet light and radiations from the sun and outer space called cosmic radiation, and by radiation from x-rays and from nuclear fission reactions and their by-products. Mutations can also be caused by an ever-increasing list of chemicals. The general name for any agent that causes mutations is *mutagen*.

There are two outward effects of all these mutagens on

living creatures: effects on reproductive cells, and effects on body cells. Mutations affecting the DNA of reproductive cells of the ovary or testis damage the DNA that is about to be passed on to the next generation. Obviously, such damage is especially serious. Mutations in the DNA of all other body cells will probably go unnoticed, although they may rarely, after long periods, result in cancer. Mutations in bacterial cells have the same effect as mutations in reproductive cells, because bacteria are single cells that divide up their DNA and pass it directly along to their offspring during cell division.

Mutagens have a variety of effects on DNA. They can make adjoining nucleotides stick to each other, and they can widen the space between a couple of nucleotides. But two different kinds of mutations are of most interest to us. In the first, one normal nucleotide is replaced by a different, but normal, nucleotide:

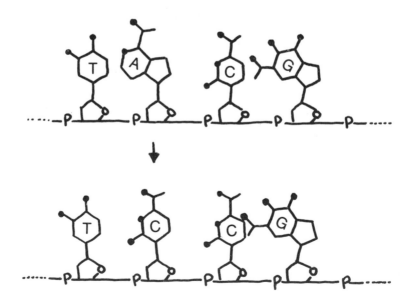

In the second, a normal nucleotide is either removed from or added to the chain:

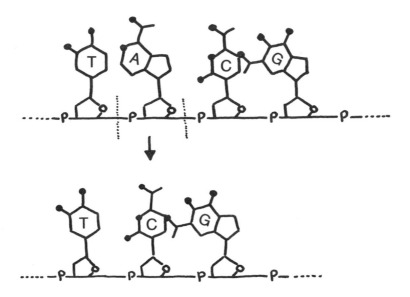

Many of these changes in DNA are propagated and perpetuated when DNA is duplicated, because they tend to cause mistakes in the copying process. In the end, the altered DNA causes defective proteins to be made by mechanisms we shall be examining in this book.

We'd be in serious trouble if our DNA had no way to protect itself against the depredations of radiation and mutagenic chemicals, but DNA doesn't take these insults lying down. DNA can *repair* changes made by mutations! Special enzymes are constantly monitoring DNA molecules to search for any irregularities. When altered nucleotides or stuck-together nucleotides or extra nucleotides or missing nucleotides are found, enzymes open the chain, excise the faulty section, add the right nucleotides, and then close the chain again. How we are thus guarded and protected by our own cellular machinery depends on the structure of DNA, which we will examine later in some detail.

*

Molecules made up of small units linked together in long chains are called *polymers.* DNA is a polymer, and its units are nucleotides of four kinds. Proteins are also polymers. Their units are called amino acids. There are exactly 20 different amino acids in proteins. It is remarkable that *all forms of life on this earth, from humblest to most arrogant, simplest to most complex, are composed of DNA and proteins made of the same four nucleotides and the same amino acids.*

In DNA the four nucleotides are different, but the chemical linkage between them is identical. It is the same with proteins. The amino acids are different, but the linkage between them is the same. Each amino acid displays a different chemical grouping that determines its unique functional role when it is in a protein. Some amino acids are water-rejecting; others seek out water. Some are acid, some are alkaline, some cause the chain to bend, some bind readily with other proteins or molecules.

There is thus a whole world of chemical reactivity in a protein molecule determined by the amino acids it contains. Because of its great chemical heterogeneity, much of a protein's affinity and reactivity is with itself. Its various chemical groupings make it fold, twist, and wrap into a very unique three-dimensional structure. Thus, a protein's shape and special function are determined by which amino acids are in its chain and by their locations on the chain. One way to visualize a protein is to imagine a long chain with a lot of hooks and sticky protuberances along its length. If you bunch up the chain, the hooks attach to different parts of the chain and to other hooks, the sticky parts adhere to other parts, and you get a permanent three-dimensional shape. This shape depends on the number and location of the hooks and sticky knobs. Proteins in their normal state have three-dimensional shapes they naturally assume because of the kind of sequence of the amino acids they contain. Unlike the chain with hooks, the linkages in proteins are exact and carefully worked out over billions of years of trial and error. Many proteins are rather globular (like a tangled ball of string) and have special locations in them that facilitate chemical reactions. These are enzymes. Hemoglobin, with its special function of binding oxygen in the lungs and releasing it in the tissues, is rather like that. It is two pairs of proteins working together, each with an iron atom (Fe) in it, that is uniquely able to bind and then give up oxygen.

Other proteins form long fibers, and many of them relate to each other in parallel rows such that they slide past each other, causing a shortening of the bundle. These are muscle proteins.

All these complex protein configurations can be chemically treated to convert them to one long, stretched-out chain of amino acids in which all of the chemical groupings are hanging free.

This opening up of a natural chain without breakage is

called *denaturation*. If natural conditions are restored, the chains will often spontaneously reorganize themselves into their complex folded pattern. This is very strong evidence that the number and the order of amino acids in the chains are the only determinants of a protein's final structure.

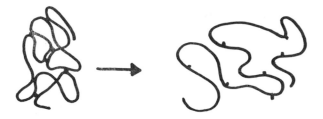

The specific and unique qualities of all the forms of life we know are determined by proteins. Proteins are the machinery of life, as enzymes, and much of the structure of cells as well. The shape and activity of every protein in the body is exactly, specifically, and uniquely adapted through years of evolutionary trial and error to the performance of particular jobs. Nothing else will substitute. Loss of a protein through damage to its gene produces an unambiguous loss of function.

❊

Science searches constantly for unifying simplifications. For those who derive aesthetic enjoyment from the explorations of science, the advance of genetics is particularly rewarding. To gaze on beauty is good; to learn of the plan behind it is better. We devote this and the next chapter to genetics, which played a major part in generating the revelations of molecular biology.

We begin with the genetic study of a human disease. Nowhere has science had a more exalting impact on the human condition than in medicine. Before medicine established its enduring partnership with science, the physician had little to offer those who were suffering. As often as not, he compounded nature's ravages. With the growth of science-based diagnosis and treatment, the patients' chances of self-healing with nature's help improved dramatically. A half century of painstaking work by scientists to understand infectious disease processes, leading to immune and antibiotic aid, freed humans of another heavy burden. X-rays, drugs, anesthesia, and improvements in surgery and nutrition have continued to advance.

With the investigations that led to understanding the disease sickle cell anemia, medicine made a payment on its debt to science. This tragic disease was first described by James B. Herrick, a physician, in 1910. It is a severe anemia peculiar to black people. The disease begins early in life, is painful, and is often fatal. The red blood cells, normally shaped like doughnuts with the center indented instead of punched out, are grossly distorted in shape in patients with this disease. In the blood returning to the heart from the tissues (blood whose red cells are depleted of oxygen), the cells are shaped like crescents, or sickles. This makes them stick to the walls of capillaries and block the flow of blood. The red cells are also more fragile than normal and break up in the capillaries, releasing the vital oxygen-carrying hemoglobin. This loss of hemoglobin is called anemia. After blood passes through the lung, where the hemoglobin in the red cells picks up oxygen, the cells resume the normal shape. Arterial blood — blood rich in oxygen on its way to fuel the food-consuming combustions in all the body's cells — has normal-appearing red cells.

Sickle cell anemia is inherited; parents pass it on to their children. There is also another form of the disease. Some people walk about without symptoms but have red cells that can be made to sickle when exposed to low oxygen *outside the body,* even though they behave normally in the body. This is called the *carrier* state for reasons soon to be apparent. Carriers can pass on their carrier state to their children, and if they have children by another carrier they can produce the full-blown disease in their offspring.

The great chemist Linus Pauling, of the California Institute of Technology, and a Harvard Medical School professor and physician named William Castle, a specialist in blood diseases, were together at a meeting in 1945 in New York. Castle told the gathering of his interest in sickle cell anemia and described the blood condition, including the fact that oxygen deprivation caused the cells to assume a sickle-like shape. Pauling, a man generously endowed with a powerful curiosity and an exuberant joy in the search, listened to Castle with growing excitement. Pauling perceived immediately that the sickling phenomenon must be due to a defect in the hemoglobin molecules inside the red cells. Pauling was primed for that interpretation. Probably no one in the world knew more chemistry, particularly the chemistry of protein molecules, and most particularly the chemistry of the proteins called hemoglobin. He returned to California after that meeting with sickle cell anemia as a new inspiration.

Pauling was the undisputed leader of the *structural* school of the new molecular thrust in biology. The structuralists were the physicists and chemists in America and Britain who were moving in increasing numbers into biology. They sought to elucidate biological function by analyzing the detailed structure of large molecules, especially proteins and DNA. They differed markedly in background and technical approach from those who used genetics or biochemistry to approach similar objectives. Pauling had already completed a large body of work on the structure of small molecules and the nature of the forces that hold molecules together that was to bring him the Nobel Prize for chemistry in 1954. (In 1962 he won a second Nobel Prize for his work in effecting a treaty to ban the testing of atomic explosives in the atmosphere. The awarding committee was probably even more influenced by Pauling's courageous espousal of unpopular causes in a McCarthy-paralyzed America of the early 1950s.)

Pauling, as I said, knew as much as anyone about hemoglobin. He knew that when the body makes its red cells, it packs them full of hemoglobin. Hemoglobin, with the help of iron atoms in its middle, picks up oxygen as the red cells tumble through the capillaries of the lungs and delivers the oxygen through the capillary walls when the red cells get to the body's tissues. Pauling knew from experiments in his lab that attachment of oxygen to hemoglobin slightly changes the shape of the molecule. He reasoned that if the red cells of persons with sickle cell anemia change their shape, especially in low oxygen, it might well be due to a mass change in shape of the many hemoglobin molecules inside the cells.

But Pauling marveled most at the intriguing fact that all the patients' trouble was brought about by a change in a gene. As far as it was then known, the genetic change in sickle cell anemia was a simple change involving a single gene. And so it was that *he saw the possibility of linking a single gene change to a precise change in a single protein molecule.*

Pauling and his younger colleague, Harvey Itano, plunged

into the problem of discovering the chemical difference between normal hemoglobin molecules and sickle cell hemoglobin molecules. What was the exact nature of the wound in the protein? After three years of intensive work, they surfaced in 1949 with proof that sickle cell hemoglobin molecules are definitely different from normal hemoglobin molecules in only one way. They differ in their electrical charge. They are, then, *chemically* different, because the electrical charge of molecules is determined by their chemical composition.

While Pauling and Itano sought the chemical change in hemoglobin, a geneticist named James Van Gundia Neel, equally fascinated by this disease, made a massive study of the families afflicted with sickle cell anemia. His careful work showed that the sickle cell gene is inherited in humans just the way Mendel's pea plant genes are. Thus, humans have two genes specifying the protein hemoglobin, one from the mother (H) and one from the father (H). A patient with full-blown disease has both genes damaged (hh). The parents of such a victim may have sickle cell anemia themselves (have both hemoglobin genes damaged, too) or they may be carriers. Carriers were proved to be persons carrying one damaged gene and one normal gene (Hh). Carriers are what geneticists call "heterozygous" — the two genes governing a particular function are different. In contrast, in "homozygous" persons the two genes are the same, either both normal (HH), or both damaged (hh). Persons who are heterozygous for the sickle cell gene are symptomatically normal, because their one good gene makes enough normal hemoglobin to keep the cells from sickling in the body. Their carrier state can be detected by taking some blood, exposing the cells to low oxygen, thereby causing enough distortion of hemoglobin molecules to bring out the sickling trait in the cells.

The sickle cell carrier state illustrates the generality that if one gene of a gene pair is damaged by a mutation and is thereby assumed to be completely knocked out, the other gene can still command the required gene function, the making of a particular protein. In such a gene pair, the working gene is said to be *dominant;* it manifests itself by causing enough (about half) of an essential protein to be made to take care of an organism's needs. In Mendel's pea plants, for example, the smooth-pea gene of the pair is dominant because it causes enough protein to be made to accomplish the state of pea smoothness. It is not known what that protein is. The wrinkled state is only seen when *both* genes of the gene pair are damaged and consequently *no* protein, or a functionless protein, is made.

With hemoglobin the same rules apply, but the outcome is significantly different. If both genes of the hemoglobin gene pair are defective, the result is very different from the benign wrinkledness in peas. The genes cause the production of all defective hemoglobin molecules. These altered proteins are better than no hemoglobin at all, for one cannot live without hemoglobin. They are packaged into red cells by machinery that doesn't see anything wrong with them, and once in the red cells they function improperly. Most genetic diseases in humans — and many hundreds of them have been discovered during the past few decades — are not like sickle cell anemia. Instead of the defective gene's protein being hurtful to the body, the protein is simply *not there* or is inoperative. The trouble, then, is caused by the lack of an essential protein.

Pauling and Itano carried genetic exploration one important step beyond Beadle and Tatum. They provided the first evidence that a gene mutation caused a specific, identifiable defect in a protein molecule. But neither Pauling nor anyone else yet knew what wound hemoglobin had sustained when its gene was damaged by mutation. Hemoglobin's electrical properties were changed, and this could explain how the protein changed the shape of red cells. But what caused the electrical properties to change? The methods then available for analyzing the proportions of the 20 different amino acids in protein were not sensitive enough to reveal possible changes.

Pauling and Itano's discovery left little room for doubt that a gene mutation could cause damage to a protein. Another experiment done at about the same time, using bacteria instead of humans, even more clearly invited the same conclusion. Bernard Davis and Werner Maas produced bacterial mutants that could make a substance essential for their growth at one temperature but could not make it at a higher temperature. They isolated the enzyme responsible for the synthesis of the substance and found it could work in the test tube at one temperature but not at the higher temperature.

Thus, mutation-induced damage to a bacterium's function was explained by damage to an enzyme (protein), just as mutation-induced damage to a human's red blood cells was explained by damage to hemoglobin.

*

In the story of molecular biology, one laboratory that stands out for ingenuity, imagination, and excellence of performance is the Cavendish Laboratory at Cambridge, England. The Cavendish was a focal point of lively discussion and the nourishment of novel notions. Here, ideas were born and raised with the tenderest of care, and the atmosphere was thick with the excitement of the chase. The sickle cell story now moves to Cambridge. During 1955 and 1956, sickle cell hemoglobin's malady was unequivocally diagnosed by a young English scientist named Vernon Ingram. Ingram had come to the Cavendish to work on another project, but finding it dull had, at the suggestion of Francis Crick, a member of the laboratory, zeroed in on the obvious but all-important question: Did the amino acid(s) in sickle cell hemoglobin differ from those in normal hemoglobin? The question was obvious. Hemoglobin, any protein, *is* its amino acids. If something is wrong with hemoglobin, the secret must lie in those links in the chain.

Remember, the 20 amino acids are found in different total amounts and in different sequences in different proteins. A protein's amino acid sequence is rigidly fixed and invariant for each particular protein. Once the chain is built with a genetically determined sequence, the final shape and function is a foregone conclusion. Hemoglobin is made up of four folded chains. Each chain is about 140 amino acids long and, when folded up, contains an iron atom to hold oxygen.

Ingram began his work in an exceptionally stimulating and generously helpful environment that was ideal for his particular project. Frederick Sanger was nearby at Cambridge University finishing up a tour de force revealing the complete amino acid sequence of the protein insulin, for which he later received the Nobel Prize. Ingram's mentor, Max Perutz, and Perutz's associate John Kendrew, were working out the three-dimensional shape of hemoglobin using x-ray diffraction techniques. Both Perutz and Kendrew were later awarded a Nobel Prize. And Francis Crick was at Ingram's elbow, a considerate intellectual needler with perhaps the most imaginative mind in biology.

Ingram and Crick naturally assumed that the difference between normal and abnormal hemoglobin would be found in the sequence of amino acids in one of the 140-unit chains.

Ingram invented a powerful technique, much simpler than anything previously available, for determining small differences in amino acid composition of proteins. The method involved breaking the long hemoglobin chains apart into shorter lengths using enzymes that snip protein chains at very specific points. He separated the resulting shorter chains by taking advantage of the fact that they moved on sheets of filter paper at different rates in different solvents and in different electrical fields. The short chains ended up being spread out on a square sheet of filter paper and could be seen by spraying on a stain that reacted with protein. He called these patterns of spots *fingerprints,* an apt name because the array of spots was distinctly different for each protein so treated. Looking for a change in one or a few amino acids in a whole, very long protein chain is definitely like looking for a needle in a haystack. But an amino acid change in a *short* chain would be readily manifest, if it modified the chemical properties of the chain. Every time Ingram treated normal hemoglobin this way, he got the same fingerprint. When he treated sickle cell hemoglobin exactly the same way, most spots were unchanged but *one* was different. A spot he expected to appear in a certain location in normal hemoglobin appeared somewhere else in the fingerprint of sickle cell hemoglobin. This immediately meant that the difference between the two hemoglobins *was* an amino acid difference. So he focused on that one changed spot and analyzed the amino acids in it. He finally proved that *one* amino acid in the short chain was changed, and he found no other changes in the other chains. It followed that in the full length of a hemoglobin chain, only one amino acid was changed! A glutamic acid in normal hemoglobin was changed to a valine in the sickle cell hemoglobin.

position number	normal hemoglobin	sickle cell hemoglobin
1	valine	valine
2	leucine	leucine
3	threonine	threonine
4	proline	proline
5	glutamic acid	glutamic acid
6	glutamic acid	valine
7	lysine	lysine
8	serine	serine
9	alanine	alanine
|	|	|
|	|	|
|	|	|
|	|	|
|	|	|
146	histidine	histidine

Ingram's accomplishment was doubly gratifying. It showed that a mutation could wreck a very long, complex protein by changing just *one* of its units — like killing an elephant with a dart. It also showed why the damaged hemoglobin was electrically altered, because the normal glutamic acid has an electrical charge that the substitute valine lacks.

The human body is incredibly tough and resilient, but

genetic disease taxes its vulnerability to the utmost. Although we possess tens of thousands of different kinds of proteins, and each of these proteins has hundreds of amino acids in it, a switch of one amino acid in one protein can kill a man or woman with sickle cell anemia. In some areas of the world, one person out of five suffers from sickle cell anemia.* Wounded by a pin prick in a gene, their bodies sicken and suffer and die from a malady no drug can help.

Pauling and Ingram had explained the disease. Plunging deeper than Garrod and Beadle could go, they had reached a new truth. And agile minds were set to work. If a mutation in a gene could cause one amino acid to be changed in a protein, there must be some simple, direct relationship between mutation and amino acid change, between gene and protein chain. The drive to define precisely this relationship led to one of the most intense intellectual races in science.

But the first, the more urgent need, was for a better model. A human system wouldn't permit finding out what the mutation actually was in DNA, that is, how the gene was changed. A flexible system was needed in which experimental manipulations would not have Frankensteinian overtones. Moreover, large populations were needed. Bacteria and viruses, always in crowds, beckoned to the experimenter. They were to become the terrain of the new explorers.

---

* The reason for such a high incidence of a disease that might be expected to have killed off its sufferers is fascinating. The high incidence is found in areas where malaria is endemic. The malaria parasite worms its way into red blood cells. The altered shape of sickle cells impedes the entry of the parasite, thus giving those with sickle cell anemia a kind of immunity from malaria.

CHAPTER FIVE

Mutations Help to Analyze Function

The sequence hypothesis. The use of mutation and selection to elucidate living function. Genetic recombination and gene mapping.

LET US REVIEW briefly what we know about genes so far:

1. They are discrete physical entities that replicate without change.
2. They can be changed by mutation.
3. They make it possible for cells to *do* something by causing proteins to be made. The proteins are the doers.
4. When a gene changes by mutation, a protein is changed.
5. One such mutation-induced protein change is an amino acid substitution in the chain.
6. Chromosomes contain all of the cell's DNA. Genes are arrayed in a line along chromosomes. DNA can be isolated from cells, put into other cells, and change their inheritance. Genes, then, are DNA. DNA is a long chain molecule, so perhaps genes are linear sections or sublengths of DNA.
7. Because living creatures perform many thousands of distinct biochemical functions, they must have at least that many proteins and, therefore, at least that many genes.

These features of the gene-protein relationship had led scientists to a hypothesis. This *sequence hypothesis* related the DNA molecule to the protein molecule in a simple and specific way. The logic was straightforward. Proteins are chains of amino acids, and genes are chains of nucleotides, so there must be some simple linear relationship between amino acids in proteins and nucleotides in DNA.

Many of the best minds in molecular biology were using this as a working hypothesis from 1950 through 1965. Crick formalized the idea in 1957. The question became, simply, *How is the sequence of amino acids in proteins determined by the sequence of nucleotides in DNA?*

Thanks to the ingenious techniques pioneered by Fred Sanger at Cambridge University, amino acid sequences of an increasing number of proteins were being worked out successfully in many laboratories. The methods were similar to those used by Ingram. Stretch out a protein chain, break the chain at different links with specific enzymes, and then, by matching up the resulting fragments, deduce the sequence. This works well for proteins, but still did not allow for discovering the order of nucleotides in DNA. The only means of getting nucleotide sequence information on DNA was indirect. It was called gene mapping, and depended on a phenomenon called *recombination*. What was needed for recombination studies were large numbers of mutant genes. Before describing gene mapping, then, let's look at how to get access to these mutant genes.

The methods of the bacterial and phage geneticists were refined from those used by Morgan on fruit flies. But what Morgan could do with fruit flies was crude in the extreme compared to the analytical power of the modern geneticists with their phages and bacteria. Part of the reason for the greater precision was that the genes of bacteria weren't double. You will recall that in Mendel's plants and, indeed, in all higher living forms including fruit flies and humans, two genes (one from each parent) represent each function. These two genes reign over their function or trait — a protein — for the life of the organism. Well, bacteria don't have parents. They duplicate their chromosome, and thereby each gene set, before cell division; when they divide, they deliver one duplicate set of genes to each daughter cell. This single-gene-for-each-protein state of affairs means that when genes are altered by mutation, the effect on function is immediately manifest. There are no recessive genes to mask results.

In addition, as we learned earlier, bacteria give the experimenter unrestricted access to enormous numbers of identical cells. The experimental tools are the same as those used by evolution: *mutation* and *selection*. Both can be used at will and with precision. Bacteria are exposed to radiation to damage their genes, and then selection pressures are applied to isolate from the total irradiated population the particular damaged organisms one wants. The power of this technique lies in the fact that *extremely rare* mutants can be detected.

Mutational analysis is analysis of function by elimination. A grisly analogy would be my trying to discover what each scientist (gene) in my institution does by pumping a pathogenic virus into the air conditioners. Some scientists will get sick enough to stop work, and I'd learn what they did by detecting the absence of publications in a certain field and the accumulation of materials the scientist was wont to use!

Some of my readers may spot a contradiction here. We

have repeatedly alluded to the damaging effect of mutations, which produce disease in humans and now biochemical deficiencies in bacteria. But we have also noted that
mutation is the basis for introducing the variation in organisms on which evolutionary survival depends. Is the damaging effect of mutation in conflict with mutation-induced
evolutionary survival? It is not. Mutations produce altered
proteins, which in turn cause a change in some function.
Inevitably, *most* such changes will be deleterious. After 3
billion years, an organism is a symphony of harmoniously
integrated functions (proteins). A random change in a note
is not likely to improve the composition. Yet in rare circumstances, a changed note may significantly improve the music, and in evolution the billions of years available give
ample opportunity for many "rare" events to take place.
Thus it is that genetic change and biological betterment
can occasionally have the same physical basis.

Let's consider a real experimental example. Bacteria (in
this case, *Escherichia coli,* a common intestinal organism)
are able to make all of the compounds they need for growth:
the 20 amino acids, the four nucleotides for DNA, vitamins,
and so on. Suppose we want to learn how a bacterium
makes a particular amino acid, say valine.

Knowing an organism can make valine is not enough for
us now. We are probing deeper. We want to know what
proteins (specifically, what enzymes) are involved in the
synthesis of that amino acid. What are the proteins whose
synthesis is controlled by the genes controlling the trait of
valine-making? We proceed by taking a large number of
normal bacterial cells (say, 10 million), all identical (grown
from one cell), and able to make valine, and exposing them
to a mutagenic agent. X-rays, ultraviolet light, or a chemical

mutagen will do the trick. Using a sufficiently high dose to produce extensive damage to many genes, we will kill most of the cells. Some cells, however, will remain alive, and some of these living cells will have damaged genes. A few of these can be expected on a purely statistical basis to be damaged in the particular function we're interested in, the making of valine. Cells with this particular defect will be rare among the millions, so must now be *selected* from the mass of remaining living cells that may still be normal or may have other problems we're not interested in.

The selection process is where the ingenuity of the experimenter is really challenged. One widely used selection technique takes advantage of the known fact that penicillin kills cells only when they are growing. After cells have been treated with a mutagen, they are put in a medium with penicillin containing all they need to grow *except valine.* In this liquid medium, most cells begin to grow because they can make their own valine; they are quickly killed by the penicillin. The rare cells that can no longer make their own valine do not grow and so are spared killing by penicillin. A centrifuge is used to drive the cells to the bottom of their container, the penicillin medium is poured off, and the cells are resuspended and spread out on a gelatin plate containing medium *with valine* and *without penicillin.* The surviving cells, now supplied with valine, multiply rapidly and their colonies can soon be harvested. We now have a plentiful supply of mutant cells selected for their inability to make valine. They breed true and are ready to be subjected to biochemical studies in the search of the exact damage to enzymes.

Bernard D. Davis, the inventor of this method in 1948, was slow in getting around to publishing it. One day Joshua

Lederberg — of whom we will hear much more later — came to visit Davis. He said that he'd learned from Salvador Luria that Davis was using penicillin to select bacterial mutants. Lederberg had independently done exactly the same experiment and had just sent off the manuscript to the publisher. He asked the journal to delay publishing it if Davis would like to write up his experiment so that the two articles could be printed in the same issue. Davis agreed, and so it was done.

Another very useful and quite different selection method invented by Lederberg is called *replica-plating*. From a mutagen-treated population of bacteria growing in liquid you can take a drop and spread it out evenly on a gelatin-covered dish containing nutrients. If there aren't too many bacteria, you'll get a dish looking like this after you incubate it for two or three days:

Each little mound is a clone or colony of a few million cells derived from one cell that landed on that spot when you spread the liquid over the surface. Now you take two more dishes. One contains gelatin with a medium rich in everything bacteria need to grow. The other contains gelatin and a medium made of only sugar and salts, the kind on which only healthy, unmutated, normal cells can grow. Next you take a round piece of velveteen the same size as the dishes and press it down gently and evenly on the colonies. Lift it and press it down, in turn, on the two new dishes. You've made exact replicas of the colony distribution of the original dish on the two new dishes. Incubate those two dishes for a while longer to allow the cells to grow. You've made a replica print, like a block print, of the original dish, with a small difference.

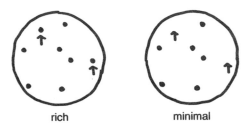

rich   minimal

Normal cells, the vast majority of the mutagen survivors, will grow up on both of the dishes, because they can make everything they need to multiply. So the pattern of distribution of their colonies will be a perfect replica of the pattern of the original dish.

Mutant cells that can't make an amino acid would appear as colonies on the rich medium but wouldn't be able to grow on the minimal medium. In the final step you'd look for rare colonies that were on the rich plate and *not* on the salt and sugar medium. You could pick those colonies off the rich plate with a wire scoop, put them in liquid medium with all they need for growth, and produce a pure culture of the mutant cells.

The beauty of the many selection methods of microbiology is that they allow you to pick up just exactly the mutants you want. Thousands of other damaged cells are not seen because they're not sought. The other charm of the technique is that each time you do an experiment you are simulating, in the laboratory, the phenomena of evolution: genetic change followed by selection of those able to survive in the particular environment you design.

What do we want with mutant organisms that can't make valine? First, they are essential to developing a full understanding of the cell's enzyme machinery for making valine. Second, they are essential for gene mapping, that is, for locating genes on the chromosome. Let's briefly examine how the cell assembly lines for producing valine and other essential building blocks have been brought to light.

If some starting material A must be converted to a final product X (such as valine), several steps are probably involved, and each step would require an enzyme. Let's suppose there are three steps:

$$A \xrightarrow{\text{enzyme 1}} B \xrightarrow{\text{enzyme 2}} C \xrightarrow{\text{enzyme 3}} X$$

If enzyme 2 is knocked out by mutation, two things happen that can easily be measured. First, X is no longer made and the cells cannot grow without being fed some X. In addition, B accumulates in the cell. Enzyme 1 is still converting A to B, but because enzyme 2 can't use it, B piles up in the cell and can be detected chemically. To the biochemist, B is a signpost to a problem with enzyme 2. He can extract from normal cells an enzyme that will convert B to C and thereby unequivocally identify enzyme 2. In an automobile production line where workers add parts one at a time, the failure of one worker to attach bolts would cause the production of the finished car to stop and cause an accumulation of bolts beside the recalcitrant worker. The pile of bolts could lead an investigator to him. So it is that ingenious biochemical techniques have discovered all the enzymes in the many pathways of synthesis of amino acids and nucleo-tides and have characterized in detail the chemical trans-actions involved. It has been found, for example, that it takes about 70 enzymes to make the 20 amino acids, starting from simple substances.

Now we come to mapping, the art of determining the location of genes on the chromosome. Once the geneticist has obtained several mutants damaged in different func-tions, it is possible to determine the relative linear position of the damaged genes on the organism's chromosome. Gene mapping, invented by Morgan to determine the order of genes on the fruit fly chromosome, is based on a phenome-non called *recombination*.

In bacteria, which do occasionally mate sexually, as we shall soon learn, the chromosomes from mating parents pair up longitudinally and then recombine, that is, they ex-change segments along their lengths. The two mixed chro-mosomes then separate and are apportioned to each of the daughter cells.

One now has a pair of cells, each containing genes from *both* of the parents all linked linearly in a single chromo-some. (We noted earlier that bacteria have only one chro-mosome, a single chain of genes. They are said to be "hap-loid," containing one copy of every gene. Higher organisms are "diploid," having two copies of every gene.)

Let us suppose that we prepare by mutagenesis and selection two mutant cell populations, one that can't make valine and the other that can't make another amino acid, phenylalanine. Each population, then, has a different dam-aged gene and therefore a different damaged enzyme. We now mix together a number of each type and allow sexual mating and recombination of genes to occur. After allowing

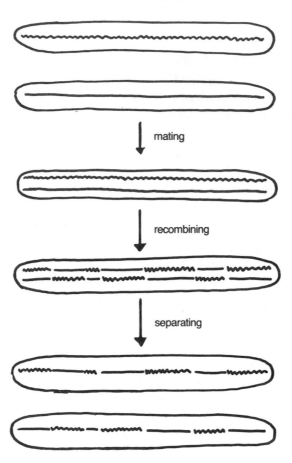

time for progeny of the union to grow by repeated cell division, we spread the organisms out on a medium lacking both valine and phenylalanine and look for normal colonies, that is, clones that have regained, by gene recombination, the ability to make these amino acids themselves. *The frequency with which normal recombinant colonies appear as a result of the union of two populations of organisms each carrying different gene defects is a measure of the distance between the genes.* This is a restatement of the principle

Morgan discovered in fruit flies, which we described in Chapter I.

This principle of gene mapping is important to much of our story and I shall restate it in a different way. The chromosome (or genome) is a long thin chain along which all of the organism's genes are strung out as lengths of linked nucleotides. The average gene is about 1,000 nucleotides, or links, long. Somewhere along the chromosome length is a gene controlling valine synthesis, and at another location is the gene for phenylalanine synthesis.

The experimenter wishes to know how far apart these two genes are (the distance d). To determine this, he produces two mutant organisms. One has a mutation in the valine gene:

The other has a mutation in the phenylalanine gene:

He then mates these two organisms (many millions are mated simultaneously, of course). In the mating process, the respective chromosomes line up in parallel:

In all the mated organisms, many random exchanges of lengths of chains occur along the full length of the chromosome, including *between* the two genes we are interested in.

This phenomenon of chains unlinking and relinking to the opposite chain is called *recombination*. Thus, two new chromosomes containing lengths of each other are created. If some exchanges occur between the valine and the phenylalanine genes, we will have some chromosomes that contain *both* genes mutated:

We will also have some chromosomes with both genes normal:

In the large population of organisms, the fate of most organisms is not of interest to us. We are interested only in the recombinants with the normal chromosome. These cells can easily be grown in the medium in which the parent mutants and the double mutants can't grow, and we end up with a pure culture of normal recombinant cells.

Because we are dealing with large numbers of organisms, and because recombinational events occur randomly anywhere along the two chains, the number of normal recombinants produced by the mating will be greater with greater distances between the two genes. Stated otherwise, the closer two genes are to one another, the less likely it is that a chain linking-and-rejoining event could occur between them. This is the physical basis for the technique of gene mapping. By damaging genes by mutation, then causing matings between pairs of mutants and counting the number of normal recombinants arising, we can locate genes relative to each other on the chromosome.

It may be enlightening to compare two chromosomes undergoing recombination with two freight trains exchanging box cars. Let's suppose that each box car (gene) in each train (chromosome) is loaded with different items and that the order of loading in the two trains is identical. That is, box car no. 1 in both trains carries television sets, box car no. 2 in both trains carries plumbing fixtures, and so on.

In each train one box car has been robbed of all its contents (mutation). In train A, box car no. 15 has been cleaned out of its contents of men's clothing. In train B, box car no. 48 has lost its consignment of kitchenware. The trains are on parallel tracks and there are many switches between the two tracks to permit cross-coupling of the two trains across the tracks. The object of the exercise is to make a new complete train with full box cars no. 15 and no. 48. This can be accomplished by connecting train A with train B across any switch that is between box car no. 15 and box car no. 48. Once the box cars are recoupled, a fully loaded train can move away and successfully complete its mission of supply.

It should be apparent that the farther apart the defective box cars are, the greater the number of opportunities there will be to turn out fully loaded trains. Thus, *distance* along the track is directly related to the frequency of switching opportunities.

If we were dealing with hundreds of such two-parallel-train combinations, each with pairs of empty box cars, the *frequency* with which repaired single trains would be generated from pairs of robbed trains would be proportional to the distance between the empty box cars. Put another way, reparative switching would happen more often when the empty box cars were some distance apart.

Gene recombination is the normal process of gene exchange in all living organisms. It is the means by which genes from our mother and our father, which we each carry in all our cells, are shuffled into a single chromosome in sperm and egg. Such gene exchange insures genetic variety.

Recombination is not to be confused with recombinant DNA technology or gene splicing, now so much in the public eye. This is an artificial laboratory device for recombining (splicing) useful genes onto bacterial DNA, returning the recombined DNA to the bacteria, and allowing the bacteria to multiply the DNA during their own growth. Large amounts of selected genes are thus produced. Again, we depend on the rapidity of bacterial growth, the large populations that can be produced, and our great familiarity with this organism. Gene splicing has important implications for the production of large quantities of important genes, and for the study of gene structure and control, but is not within the scope of our story, the story of scientists unlocking the mystery of what genes actually are.

CHAPTER SIX

Sex in Bacteria

The evolutionary meaning of sex. Joshua Lederberg discovers bacterial conjugation. William Hayes shows how conjugation works.

IN THE EARLIEST WHISPERINGS of life on earth 3 to 4 billion years ago, primitive cell populations had little opportunity to survive unless they could adapt swiftly and, in general, use variation and diversity to escape extinction. Variety for them was not the spice of life, it was the essence of life. Heterogeneity in living systems is created and sustained by continual changes in DNA, produced by mutations. Mutations are chance, haphazard events, more likely to be destructive than constructive. But they make it possible to begin and continue the ever-spreading diversification that eventually produced the innumerable species of our planet. An advantageous, felicitous change in an organism's DNA gave it a slightly better chance to cope with its environment and so to reproduce its kind slightly faster. But evolution by mutation alone would inevitably be not only chancy, but extremely slow, and would be restricted to a single line of descent. Some means by which genes could also be exchanged *between individuals* would enormously amplify and speed up the thrust toward diversity. Such a process of gene transfer between cells would have been a primitive form of sex. Some form of it must have emerged early in evolution.

Perhaps the first leaning toward sex came when two cells, randomly meeting, fused to form a single large cell encompassing the contents of both. Such fusions of cells can be observed in the laboratory today. It is a process entirely lacking in discrimination. Any cell will do, and the process we know in the lab must be induced by a third party — a virus. And cell fusion isn't the real thing, for there's no pairing and recombination of the participating chromosomes. When the fused cell divides, it partitions all the enclosed DNA from both "parents" into identical progeny. There is no mixing of DNAs to produce *new* combinations.

The next evolutionary step toward real sex probably involved the introduction of specificity into a meeting of cells. This new type of meeting would not have been a random fusion on chance contact, but a cell union in which one partner was chemically complementary to another. The contact surface of one partner would have to be capable of binding the contact surface of the other through some kind of cellular recognition. In such associations lay the beginnings of maleness and femaleness. In conjunction with surface recognition would have come a commitment to a division of labor in the gene-exchange job. One type of cell would have given DNA, and the other would have received DNA. Donor and recipient, male and female. No matter how few genes got transferred from one cell to another by such a process, subsequent recombinations of genes inside the recipient would insure that some genes from the donor were included in the chromosomes of the offspring of the union. Gene transfer is an effective promoter of diversity because the DNA transferred is from another living organism and therefore contains genes of proven value. This contrasts with the unpredictable, chance nature of mutation.

Speculation on the origin of sex is given substance by the remarkable discovery of sex in bacteria, which is the subject of this chapter. That discovery led to the new and powerful science of bacterial genetics.

The discoverer of sexual gene transfer in bacteria was Joshua Lederberg. He is one of the giant intellects of microbial genetics, and for his work he won the Nobel Prize in 1958. He is now President of Rockefeller University.

Lederberg was a 19-year-old medical student in 1945 when he read Avery's paper on pneumococcal transformation. It changed his life. Only a year after reading the paper, he became a Ph.D. student of Edward Tatum, and made his greatest discovery. It was directly inspired by Avery's studies.

We should make it clear at this point that in the transformation process that Avery studied, two things had to happen to insure success. One, of course, was that the pure DNA added to living pneumococcal cells had to get into those cells. The second was that the added DNA had to pair up with and recombine with the pneumococcal DNA. Once this event occurred, the genetic information in the added DNA (in this case, the ability to make a polysaccharide coat so as to express virulence) could be expressed in the progeny arising by cell division of the recipient pneumococcus.

Avery's transformation experiments insistently pressed Lederberg to ask: If DNA prepared by a human experimenter could get into bacteria and recombine, must not bacteria have some mechanism for recombining their genes in nature? He considered the possibility that bacteria were like higher organisms, bringing their genomes together to effect recombination. But thinking more directly along transformation lines, he wondered whether bacteria might release DNA; perhaps this, drifting in the medium, would be taken up by other bacteria. Such a transfer of DNA between bacteria could be of great significance if it resulted in recombination. This would be detected as a genetic change in the offspring of the participants in the encounter, specifically, the appearance of genes in the offspring of the DNA recipient.

The experiment he set up to test the possibility was simple and cleanly reasoned. He was, of course, aware of the usual way bacteria propagate themselves: they divide in two to become their own children. If sexual recombination did occur at all, it would be expected to be very rare. It occurred to him at first to mix some bacteria that were mutant in one gene together with bacteria that were mutant in another gene and then look for rare normal recombinants. Such recombinants would indicate that recombination between the two genomes had occurred (as we saw in Chapter V). But he realized that this argument was flawed. Single mutations produced by mutagens like the ones we've discussed are usually the result of single base changes in DNA. This means they have a finite chance of *reverting* back to normal — the wrong base being replaced by the right one again — spontaneously. The mutation is said to be reversed, back-mutated, or corrected. That happens rarely, but when you're looking for a rare event anyway, another rare event of the kind you don't want could interfere with your experiment. You can see that if *one* of the putative parents reverts to normal spontaneously, it could not be distinguished in this experiment from a recombinant.

So Lederberg had to return to the drawing board. He had to get around this problem of back mutation. His solution was ingenious: he would use the same principle of mixing mutants and looking for recombinant progeny but would use *double* mutants. If each parent was defective in *two different* genes, the chance of either parent spontaneously reverting to normal would be vanishingly small. The reasoning is not complicated: the chance of one mutation reverting to normal is about one in a million; the chance of *two* mutations reverting to normal is one in a million *times* one in a million, or once in a trillion! He might then confidently expect to detect even a very rare recombination event (about one in a million) because it would be likely to occur more frequently than any reversion.

In our analogy of the freight trains on parallel switching tracks, one freight train with one empty box car might rarely be refilled with merchandise by a repentant robber (reversion). This would produce a normal, fully loaded train without switching. However, if *two* box cars were emptied in a train, the chance of their both being replenished by criminal benevolence would be much smaller. Under these circumstances, in which each train had two empty cars, the appearance of a fully loaded train after switching could be assumed to be the result of switching.

Now Lederberg was ready. On the next page is a schematic picture of his famous experiment.

He mixed 100 million cells of a population defective in genes A and B together with 100 million cells of a population defective in genes C and D. One strain may be characterized by the gene make-up $A^-B^-C^+D^+$, which means

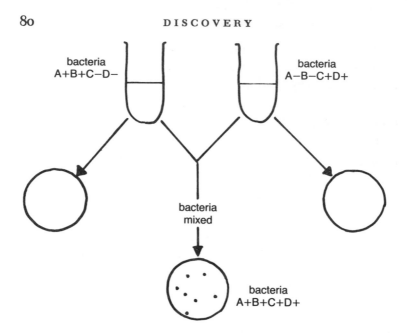

genes A and B are defective and genes C and D are normal. This strain could only grow when given compounds a and b. The other strain is of gene make-up $A^+B^+C^-D^-$: genes A and B are normal and allow the organism to make normal products a and b. Genes C and D are defective so that the organism cannot make products c and d; they must have c and d in their medium to grow. Of course, neither strain could grow on minimal medium (medium without needed factors a, b, c, and d).

Lederberg was successful so quickly that we can't build up the tension of the search! After mixing the two strains together and allowing a reasonable time for mating to occur, he spread a sample out on a gelatin dish of medium lacking compounds a, b, c, and d. Nothing but normal cells could grow on this medium. Some colonies did grow! They proved to be normal, of gene make-up $A^+B^+C^+D^+$, and so could only have been the result of an exchange of genes between the two double-mutant strains of bacteria he put into the experiment. In a vast sea of organisms thought to procreate by simply splitting in half, the powerful selection technique for finding rare recombinants of nutritionally ex-acting mutants had picked up a few organisms that propa-gated by a wholly different technique: exchanging genes!

You will note from the figure that Lederberg did not forget the essential control experiment. He took a large number of cells of each strain and simultaneously spread them on dishes *separately* to show that no normal recom-binants appeared.

Lederberg's experiment, like Luria's, was clean, simple, and conclusive: bacteria could recombine their DNAs, could exchange genes. The next question was how.

How did the bacteria manage to get their DNAs together? Was it, as Lederberg had guessed at the outset, that DNA released by some bacteria was taken up by other bacteria? But the fluid surrounding either of the cells failed to trans-form the other. Was the DNA perhaps very unstable? Or did the cells actually merge in some way and in this inti-macy share their DNA? Bernard Davis, Professor of Bac-terial Physiology at Harvard Medical School and inventor of the penicillin method for isolating mutants that we dis-cussed in the last chapter, learned of Lederberg's results and thought he knew how to answer the question. Shortly before, he'd been leafing through a catalogue of laboratory glassware and had noticed a glass tube that had a fine filter built across its middle. The filter would let fluid and things dissolved in fluid (like DNA) pass, but would not let bac-teria through. It hit Davis that he could bend the tube into a U-shape and put one of Lederberg's double mutants on one side and another double mutant on the other. In other words, he would repeat Lederberg's experiment exactly, except that the bacteria were prevented from contacting each other. If the experiment's positive result — the recombination of genes — was achieved by the secretion and imbibing of *free* DNA, the result in the filter tube should be the same, because the DNA could pass through the filter to reach and transform the bacteria on the other side. If, on the other hand, bacteria needed intimate contact to achieve recombination, then the filter, like a bacterial bundling board, would thwart their plan.

Davis added fluid containing the appropriate mutants to each side of the tube and forced the fluid surrounding the cells back and forth across the filter. No recombination occurred. He repeated the experiment. Still no recombination. So cells had to make contact with other cells to accomplish gene transfer.

Lederberg called the phenomenon he'd discovered bacterial *conjugation*. This sex-like process, involving the union of two bacteria and the intermingling of genes, was the gateway to bacterial genetics. Lederberg immediately put bacteria to work using the new-found conjugation process to reveal how genes were organized on the bacterial chromosome. Before Lederberg's work, bacterial genetics was a blank. After this work, bacterial genetics became one of the sharpest tools of the advancing new biology.

Lederberg's discovery of bacterial conjugation would not have been possible without the mutants having specific nutritional requirements. Such mutants had been discovered and put to use by Beadle and Tatum in their work with bread mold in the early 1940s. Others had made similar mutants of bacteria and used them successfully to identify enzymes involved in amino acid and nucleotide construction. Thus, Lederberg (now working with his wife Esther), with prodigious effort and ingenuity, was now able to exploit bacterial conjugation and the recombination techniques of classic genetics to make a detailed map of the bacterial chromosome. They showed that all of the genes of bacteria are linked to each other, apparently strung out along the length of a single chromosome. Their map of the bacterial chromosome eventually contained all of the genes known at the time.

One final point about the Lederbergs' work. They showed that, after conjugation and recombination of genes, offspring always end up with one chromosome, one genome, one string of genes. This state, remember, is called haploid. A cell carrying two homologous (equivalent) chromosomes is called diploid. When bacterial DNAs from two cells get together in one cell there is a very transient diploid state, a parallel pairing up of two chromosomes. Recombination ensues quickly and the cell then again has one chromosome, consisting of a linear mix of genes from the two "parents." This chromosome now *duplicates* so that each daughter cell receives a complete and perfect copy of the parental chromosome.

In human cells, there are *two* sets of genes (one from each parent) and they *stay* that way, that is, they remain diploid. Cells in ovary and testis carry out a bacterium-like recombination between parental chromosomes so that a similar mother-father shuffling of genes is accomplished. Then each sperm- and egg-making cell undergoes a special division that puts one mother-father mixed chromosome of each pair into each sperm or egg. That's how sperms and eggs become haploid. They must be haploid at that stage so that when they combine to make a new individual, the normal diploid state will be restored.

The way Lederberg went about discovering bacterial conjugation is unusual in science. It is not common for a

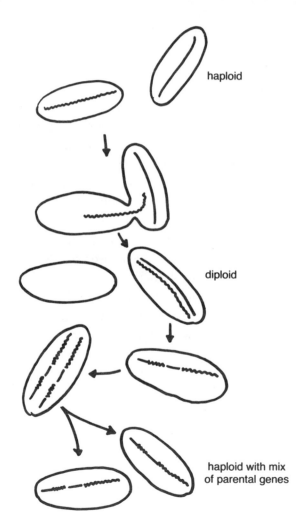

haploid

diploid

haploid with mix
of parental genes

scientist to declare his objective and proceed directly to it. The course of discovery is more often circuitous, interrupted, sometimes obscure. Unexpected occurrences and pleasant and unpleasant surprises are the norm. The scientist gets used to being led down false paths, to having the significant observation emerge while pursuing something else. The abler scientist knows how to exploit the unexpected and turn surprise to his own advantage.

While Lederberg used conjugation as a tool of bacterial genetics, he made unwarranted assumptions about how the conjugation process actually worked. Convinced that the bacteria met and shared their genomes equally as higher animals do, he stubbornly held to that view as evidence to the contrary accumulated. But Lederberg's achievement was magnificent. He not only opened the door to the analysis of the bacterial genome, he discovered an exciting natural phenomenon that is a model for an early evolutionary form of sex.

The baton of bacterial sexuality was next taken by William Hayes, a microbiologist in London. His experience was to be the opposite of Lederberg's: rich in serendipity. (Serendipity is finding one thing when you're looking for something else.) In 1952, following Lederberg's lead, Hayes was trying to learn more about the nature of the mating process. In particular, he wanted to clarify its timing: what happened from the moment two bacteria met until the mating was accomplished?

Using Lederberg's methods, he mixed the two participating mutants in liquid medium and then at regular intervals took samples and spread them out on gelatin dishes to measure the number of recombinants. This gave him a measure of the number of matings that had occurred as a func-

tion of time. Normally, bacteria are sensitive to streptomycin; this antibiotic kills them. If the gene controlling streptomycin sensitivity is altered, the organisms become resistant
to the drug. In certain of Hayes' "control" experiments,
streptomycin was present in the mating mixture. When *one*
of the parents (say, the one defective in genes A and B —
$A^-B^-C^+D^+$) was streptomycin sensitive and the other (defective in genes C and D — $A^+B^+C^-D^-$) was resistant, the
antibiotic (as we would expect) stopped the formation of
recombinant progeny. But the surprise came when the
other parent was streptomycin sensitive. Streptomycin then
had no effect on the process! That is, *one* of the parents
could effect union and recombination even in the presence
of streptomycin. Because streptomycin killed, this meant
that one of the parents didn't have to be alive for conjugation to be successful! A truly astonishing, if somewhat
macabre, discovery.

What did it mean? Hayes concluded that the two partners in the mating did not contribute equally to the union.
Contrary to Lederberg's tenaciously held conviction, the
parents were very different. Mating was proceeding between dissimilar participants. Males and females?

Hayes' insight penetrated deeper. It seemed to him likely
that the partner that could participate though dead (the
streptomycin-sensitive partner) was a *donor* of genetic material. The other partner, having to be alive (the streptomycin-
resistant partner) was the *recipient*. This made sense, because the recipient not only must receive the DNA of the
donor but must also nurture the recombination process *and*
be able subsequently to divide to make daughter cells.
These suppositions were soon confirmed by Hayes, and
then by others.

We shall not dwell at length on Hayes' further work, but will here mention a few of its main features to prepare us for the next events in the saga of bacterial conjugation.

Bacterial cells come in two mating types, called F^+ and F^-. The F^+ cells are DNA donors, or males. The F^- cells are DNA recipients, or females (F^- means lack of F). The F^+ state is conferred by inheritance of an F factor (a small piece of DNA). The F factor acts very much like an infectious virus in that it spreads to F^- cells when F^+ cells come in contact with F^- cells. Maleness, then, is contagious! For a cell to become a donor of its chromosome (its DNA), the F factor must become a part of the chromosome; it must be integrated into it just like another gene. During conjugation, the integrated F factor breaks, and one end becomes the point destined to enter the recipient. The tail, with part of the F factor attached to it, is the last point to enter the F^- cell.

The fertile state here described is rare in ordinary bacterial populations; it characterizes only one in a million or fewer cells. In fact, Lederberg was extraordinarily lucky in the strain of *E. coli* he worked with so intelligently; very few strains of this bacterium or of other bacteria mate at all! It was this one-in-a-million event that Lederberg had picked up in his experiments. In conjugation, or mating, the chromosome of the donor enters into the recipient linearly and at a steady rate, gene by gene, through a tube that transiently forms between the two cells. The amount of chromosome injected by the donor is variable, but is usually much less than its full length. This is because the donor chromosome breaks off long before its full length has entered the recipient. (In higher organisms, the genetic contributions of male and female are always precisely equal.)

One is prompted to ask why a very few organisms in a large population that normally propagates by cell division would carry on sexual reproduction. The answer appears to lie in what we've already said about the high value evolution puts on change and variability. Sex is used by all forms of life as a means of recombining genomes, mixing genes, so as to maximize genetic heterogeneity. The greater the variability of the genomes in a population, the greater the likelihood of survival of the species in changing and hostile environments. The better equipped the traveler in strange lands, the more likely is survival. Although sex is only a minor activity in bacteria, by circulating the F factor, it can eventually confer on a whole population an opportunity to shake up its gene make-up.

CHAPTER SEVEN

A New View of the Chromosome

François Jacob and Elie Wollman study the dynamics of the mating process. Interrupted mating reveals sequential gene transfer. The chromosome is found to be circular.

THE DISCOVERY of bacterial mating and the ensuing extensive mapping of the bacterial chromosome were exceptional achievements, particularly when we realize that only a few cells among millions in those populations indulged in the conjugal act. The prodigious feat of mapping the bacterial chromosome was possible because of the exquisite sensitivity of the selection techniques available, which allowed detection of extremely rare normal recombinants resulting from the mating of mutants. But when only very few bacteria were mating among millions, it was impossible to learn much about the mating process itself. What was actually happening at each moment from the bringing of the mating partners together to the appearance of progeny containing both parents' genes? As we saw in the last chapter, William Hayes had made a beginning in studying the process. He also opened the next door, by discovering a special strain of *E. coli* males that mated 1,000 times more frequently than the usual population of bacteria. This meant that a much larger proportion of a population of bacteria would be mating at any one time, so that the process itself could better be studied. Through Hayes' generosity the new strain, called *Hfr* for high frequency of recombination, was made available to many laboratories in Europe and America. It was put to most imaginative and dramatic use by Elie Wollman and François Jacob at the Institut Pasteur in 1955.

The experiment that Jacob and Wollman undertook was of sufficient importance and complexity that I must take some extra time to explain it. Their aim was to envision the mating scenario from the first contact between cells to the final appearance of offspring containing a trait or traits of both parents. As a first step, they wanted to know how and at what rates genes were transferred from donor to recipient. The availability of Hayes' Hfr males made it possible to answer this question. Hayes' male bacterium was normal with respect to the following relevant traits: it was streptomycin sensitive; it was able to make the amino acids threonine and leucine; and it could grow on the sugars lactose and galactose. The relevant part of the genetic map of the male looked like this:

The female cells to which the male would be mated lacked all of the male's normal propensities. They were unable to make the amino acids threonine and leucine, they couldn't use the two sugars, and they were resistant to streptomycin. The relevant part of the female bacterium's map (with x marking mutated genes) would be:

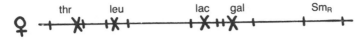

Jacob and Wollman put males and females together and immediately began removing small samples from the flask, continuing to do so for about an hour. Each sample was spread on a gelatin-covered selection dish that contained streptomycin, and lacked amino acids and sugars. On this dish, streptomycin-sensitive nonmating males among the input Hfr cells would be killed, thereby rendered unable to form colonies. The nonmating females couldn't grow out colonies either, because they couldn't make threonine and leucine. Indeed, the *only* cells that could grow up on the plate were streptomycin-resistant females that had received the male genes allowing them to make threonine and leucine and to use lactose and galactose.

They did the experiment so that new mating pairs could not form after the samples were removed. This was accomplished by the simple expedient of *diluting* the samples with fluid sufficiently so that the bacteria couldn't find each other! Any mating pairs formed before the dilution could continue the transfer of genes as long as they wanted.

Jacob and Wollman were gratified to find that recombinant colonies produced by females that had acquired male genes began to form immediately, increased steadily for about 50 minutes, and then leveled off. This meant that the first contacts between males and females were established quickly, that more and more contacts occurred until, by 50 minutes, all possible matings had occurred.

They next watched to see what *different genes* did. And here appeared a new dimension to bacterial genetics! Watching the rate of appearance of the genes permitting the making of threonine and leucine, they found that the process proceeded briskly until 18 percent of those genes from the male had entered the female (measured as recombinant progeny able to make the amino acids).

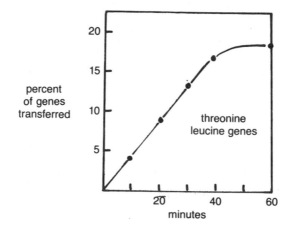

By contrast, recombinants that could use galactose appeared much more slowly. When the process stopped, only 5 percent of the male's genes had been transferred.

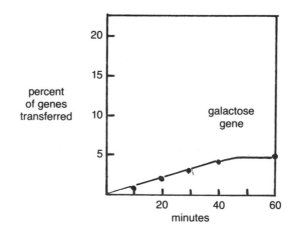

These two experiments were done identically, so it was fair to assume that the same number of male-female contacts were made in both. The difference in the results was produced by the only variable in the experiments: which gene was being followed. Jacob and Wollman inferred that the fact that the threonine-leucine genes were transferred more effectively than the galactose gene was somehow due to a difference in the effectiveness of the gene transfer process itself.

The next experiment was as bizarre in its design as it was astonishing in its result. Jacob and Wollman followed the same sequence of mixing males and females, taking and diluting samples at intervals. But this time they added one more step. They put the diluted samples into a Waring Blender briefly to *break apart any mating pairs*. They then spread out the mixture on the dishes as before. In this way, they effectively stopped continued contact between partners at different times after mating had started. In effect, they narrowed their measurement to what was happening from the beginning of mixing of males and females to the moment the sample was taken. (After the blending, no further contact was possible in the diluted samples.)

(Jacob had brought a Waring Blender back from America as a gift for his wife. Although American born, she was culinarily completely Parisian and had no use for the blender. So it came to languish in the laboratory, and Wollman conceived the idea of using it to produce what came to be lightheartedly called *coitus interruptus* in bacteria.)

The blender experiment showed clearly that genes were not transferred from the male to the female simultaneously, in a bunch. Lederberg had become convinced that conjugation involved a complete merging of male and female genes, and this new result was quite different and unexpected. Genes were transferred to the female in a regular procession, one after the other, over a period of nearly two hours! Wollman said later "the fact itself could not be anticipated. I mean, nobody had ever thought before of a chromosome which would be transferred from one end, and be injected during two hours, in bugs which normally divided every twenty minutes."

The inescapable image that then arose, and is now fleshed out and solidly established, is that of a male chromosome entering into the female at a fixed rate through a tunnel-like contact.

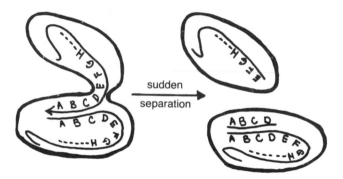

At intervals, when the mating pairs are subjected to the fierce agitation caused by a blender, they are sheared apart. The entering DNA is broken off at the contact junction, leaving within the female the length of male chromosome already transferred. This segment then pairs with the corresponding length of the female chromosome; genes are recombined, and the female then divides. *All* the males of a particular Hfr strain transfer their chromosome not only at a fixed rate, but also from the same starting point!

The excitement that these experiments kindled in Jacob

and Wollman and their colleagues at the Institut Pasteur was partly attributable to their awe and admiration of the ingenuity of *E. coli* in the management of its sex life. But even more, as is usually the case in science, it was because a brand-new vista had opened before them. An enormously powerful tool had fallen into their hands: could not the bacterial chromosome be mapped directly by the new process simply by timing the entry of genes? Why not, as before, make a variety of mutant females, defective in the genes to be mapped, mate them with healthy normal (streptomycin-sensitive) males, and time, by blender, the moments at which the earliest recombinant colonies for each of the genes appear. The time of recombinant appearance for a given gene would be a measure of the gene's distance from the chromosome's point of entry.

✽

Using the four genes in the original experiment, Wollman and Jacob found that each of them began to appear in progeny at different times. Further experiments led to their conviction that the method was valid for mapping the chromosome. They proceeded to make a detailed map.

It is important to realize that Wollman and Jacob were measuring the actual direct physical linear distance between genes along a chromosome, as with a ruler. This is quite different from Lederberg's, and classic genetics', method of estimating distances between genes along the chromosome by recombination frequency. Yet, as the results came in from the new method, they were in remarkably good agreement with the recombination methods. This enormously buoyed scientists' confidence in the value of *both* methods. The recombination frequency method, in fact, complements the interrupted mating method: the former is best for examining short segments of the genome where the genes are close together; interrupted mating is best for long distances (genes separated by at least one minute in entry time).

As we noted earlier, the F factor, which confers maleness, becomes attached to the chromosome and thereby initiates the chromosome injection process. Jacob and Wollman found that there are Hfr strains that differ only in the location of attachment of F to the chromosome. The F location point is the point of chromosome breakage. The end of the chromosome opposite the F attachment point is the end that enters the recipient first.

As Jacob and Wollman began to discover an increasing number of gene sequences that were being injected into recipient bacteria by different Hfr strains, a fascinating pattern emerged. For simplification of illustration, let's assume that a series of genes designated by letter are arranged as follows:

GHIJKLMNOPQ

As different Hfr strains were mated with females, one strain injected its chromosome in the sequence:

IJKLMNOPQGH

Another injected in the sequence:

OPQGHIJKLMN

And another in the sequence:

MNOPQGHIJKL

And another in the sequence:

HGQPONMLKJI

The glaring conclusion was that the *E. coli* chromosome was a *circle*, broken at different points around its circumference by the F factor when an Hfr strain was made! The circularity of the *E. coli* chromosome is now a solid fact of biology, having been further proved by other genetic and physicochemical studies. Indeed, it can be readily seen in the electron microscope. Like so many things revealed by science, they're remarkably easy to see once some ingenious experimenter has uncovered them.

CHAPTER EIGHT

The Structure of DNA

Erwin Chargaff seeks secret of DNA's message in nucleotide composition. Linus Pauling's model building and use of x-rays yield new insights into molecular structure. A chronicle of James D. Watson and Francis Crick's revelation of DNA structure.

THE UNVEILING of the structure of DNA in the spring of 1953 was a burst of sunlight for workers in all fields of molecular biology. They stopped in wonder and delight and saw a rich harvest before them. The revelation of structure opened up an expanding vista for the exploration of living function.

The year 1953 was the dividing line between the period of growth of bacterial genetics we have observed thus far, and the period of rapid growth of molecular genetics and molecular biology that was to follow until the logical end of one phase of it, the completion of the deciphering of the genetic code in 1964. In contrast to the sluggish reception of Avery's evidence that the chromosome was DNA, Watson and Crick's disclosure of DNA's structure was immediately received with almost universal acclaim. Part of the reason was its unexpectedness. Most workers in the field had not spent much time pondering the structure of DNA, and Watson and Crick had only begun their work a little more than a year before they hit on the solution. Then, with remarkable alacrity, scores of scientists seemed instantly to know all sorts of experiments to do next! Another reason for the instant acclaim was that the revealed structure was marvelously soul satisfying in the felicitous articulation of its components. Implicitly, in its structure, DNA revealed how it duplicated itself. Never before had a molecule's function been revealed by simply gazing upon its structure.

Watson and Crick's DNA structure and Darwin's *Origin of Species* have often been equated in terms of their impact on both science and society. Such a judgment is historically premature. We're still too close to DNA's first impact to weigh it properly. In the meantime, it is safe to say it has been the biggest event in biology *since* Darwin.

(No special significance should be attached to the order of Watson's and Crick's names in the account that follows. I usually put Watson's name first because that's the order they chose on their first publication of the structure, and it is the order most widely used. Watson tells me that he and Crick tossed a coin to see whose name would be first. He won. Both agree heartily that neither could have succeeded without the other.)

One of the reasons that the Watson and Crick search took such a relatively short time was because they already had on hand much important information needed for a solution. What they were after, of course, was the exact *three-dimensional* arrangement of the chemical parts of DNA. Their hope was that, in finding the correct shape of DNA, they'd learn how it duplicated itself and how it carried its genetic information. Their success brought an answer to the first of these hopes, but not to the second.

Neither Watson nor Crick was an experimenter. They were theorists, synthesizers, model builders. They had pieces of a puzzle before them — more pieces than they needed. Their task was to select the right pieces to complete the picture, and to reject the pieces they thought were inappropriate, misleading. They hoped in the end to hold up the finished picture and see whether it could withstand the criticism of their colleagues.

Let's see what they had going for them as they started their work in the fall of 1951. First, there was the inspiration provided by Avery seven years before. They knew that DNA was the right molecule to examine to unlock the secrets of the gene. Second, the chemistry of DNA was already well worked out. As we've noted, it always had just the four nucleotides containing the bases adenine, thymine, guanine, and cytosine. These nucleotides were known to be linked together through their phosphates and sugars, as we saw in Chapter I:

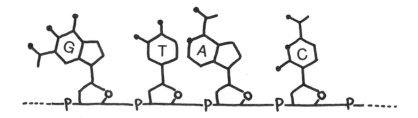

The repeating sugar-phosphate-sugar-phosphate chain is referred to as DNA's *backbone*. Third, they learned that Erwin Chargaff, of the College of Physicians and Surgeons at Columbia University in New York, had recently completed some very important work. Chargaff, like Lederberg, had been strongly influenced by Avery's transformation studies. Recall that the fact that DNA could enter bacteria and transform them genetically was the flame that ignited Lederberg. Chargaff, as a chemist, was more interested in the DNA itself. He was convinced that if DNA was the genetic material, the information for inheritance must lie in its nucleotides. In 1948 he wrote, "I saw before me in dark contours the beginning of a grammar of biology — Avery gave us the first text of a new language, or rather he showed us where to look for it. I resolved to search for this text. Consequently, I decided to relinquish all that we had been working on or to bring it to a quick conclusion — I started from the conviction that, *if different species exhibited different biological activities, there should also exist chemically demonstrable differences between* (their) *DNAs"* (italics mine). He set himself the task of analyzing the nucleotide composition of DNAs from different sources.

In 1950 he published the results of four years' work on DNA. It was the next big advance in the knowledge of DNA after Avery. He showed two things of great significance. The first was that the *amounts of the four nucleotides and their relative proportions differed considerably* in DNAs from human, pig, sheep, ox, bacteria, and yeast. Second, he discovered that no matter how much the amounts of the four nucleotides varied in the different DNAs, *the amount of adenine always equalled the amount of thymine, and the amount of guanine always equalled the amount of cytosine.* You'll not see the significance of this discovery now, and you may take comfort that Watson and Crick didn't either, for a long while.

Chargaff was certain that these nucleotide relationships must have a structural significance, but he never carried the work further. There really wasn't much more he *could* have done with his straightforward biochemical approach. Unfortunately, he was profoundly hurt and embittered by Watson and Crick's triumph three years later, feeling (incorrectly, most would agree) that their accomplishment represented a dilettantish, seat-of-the-pants, half-baked kind of science. Be that as it may, Chargaff's rules of nucleotide composition represented one of the most important sets of observations contributing to the final solution of the structure of DNA.

Another scientist who influenced the search for DNA's structure in a very different way was Linus Pauling. This uncontested world master of structural chemistry, apparently turning his attention away from protein, whose structure he had done much to reveal, was eyeing DNA at about the same time Watson and Crick were. Pauling's brilliant successes, which had long since won him a Nobel Prize, were attributed to a profound knowledge of chemistry, a fertile imagination, and a venturesomeness that early led him to the use of model building. Models are three-dimensional representations of an imagined molecular structure fashioned in cardboard, wood, or metal. The builder tries to incorporate known information from chemistry and physics into the model to get it all to fit. If all the relevant data fit comfortably, the scientist feels the model might represent the real structure in nature. That, of course, remains to be proved by arduous physical and chemical work.

Pauling made very effective use of x-ray crystallography for the study of molecular structure. He examined and interpreted the patterns of reflections made by an x-ray beam on a photographic film as it passed through crystals of substances. The same technique was being used by Max Perutz and John Kendrew, laboratory neighbors of Watson and Crick at the Cavendish Laboratory in Cambridge, England. In the early 1950s, they showed that helical chains were an essential feature of the structure of hemoglobin molecules, an achievement for which they got the Nobel Prize in chemistry in 1962. Crick, in fact, *was* a crystallographer and knew that x-ray analysis was essential to uncovering the structure of DNA. In fact, Watson had come to work with him to learn x-ray crystallography. Linus Pauling's great contributions in the use of chemistry, in the use of x-ray crystallography, and in showing the value of model building, put all structure-detectives like Watson and Crick in his debt. In addition, Pauling's dawning interest in DNA was a goad and a threat to Watson and Crick, stimulating the flow of their competitive juices.

To complete the dramatis personae, there was Maurice Wilkins at King's College in London. He was a physicist who began to work on the x-ray crystallographic structure of DNA about 1950. He was a first-rate crystallographer, and he and a young associate, Rosalind Franklin, were well positioned to beat Watson and Crick to the structure of DNA. But the Wilkins-Franklin collaboration didn't enjoy the easy rapport and open give-and-take that made the Watson-Crick style so effective. Franklin, who produced and interpreted the x-ray photographs of DNA, was a methodical, careful, cautious experimenter who did not easily share her ideas or results with her mentor. Her x-ray photographs turned out to be of great importance. When, in the course of the events we shall relate, they were seen by Crick and correctly interpreted, they came to be pivotal evidence in the great discovery.

Wilkins and Crick were good friends and they exchanged what information was available as their work progressed. Wilkins' contributions overall were important and were recognized by the award of the Nobel Prize in medicine together with Crick and Watson in 1962.

Rosalind Franklin is an enigmatic and tragic figure. She died of cancer in 1958 at the age of 37. She remains clouded by the sharp contrast between her solitary, cautious research style and the gregarious, flamboyant, and daring mode of Watson and Crick. Her studies had brought her close to the truth in 1952, and if Watson and Crick had stumbled, she might well have moved ahead of them in the race.

Watson and Crick worked together remarkably effectively during the short period of their association. Both were highly articulate, and each enjoyed uninhibitedly bouncing ideas off the other. They were perfect intellectual companions, highly critical and astute in weighing the facts. Often scientists feel obliged to include in their explanatory theories *all* data without attempting to judge their quality. Some facts, however, are harder — more likely to be true after hard testing — than others. One of the more difficult things in science is to choose between soft facts (seeming truths) and hard facts in constructing a hypothesis. Watson and Crick agreed to try to envision a structure of DNA based on as few pieces of good hard data as possible. They had to be astute enough to make the correct choices.

Watson and Crick are a fascinating pair partly because we know so much about them and how they worked, and because their personalities are so different. Watson's account of the year and a half of search and discovery in his book *The Double Helix* is spirited and astonishingly frank. It irritated and angered some people, particularly because of the insensitivity of its treatment of some colleagues, notably Rosalind Franklin, and its overemphasis on the motivating power of the competitive urge. The book's outstanding quality, however, is its candor. It sets itself apart from most accounts of scientific discovery by telling it straight, and the story is told by one who seems to be just as he was when the great events were unfolding. It should be said, however, that the story is far from representative of most scientific research in its style.

Watson was primarily a geneticist. He was a protégé of Luria and the phage group. He graduated from the University of Chicago at the age of 19 with a principal interest in ornithology. In 1950, at age 22, he got his Ph.D. with Luria at Indiana University. He embarked on his career with a fellowship to work in Copenhagen, did some uninspiring work on phage there, and then became convinced he wanted to get into molecular structure by learning about x-ray diffraction. That landed him at the Cavendish in the fall of 1951, ostensibly to work with Perutz on protein, but ending up with Crick on DNA.

Crick was ten years older than Watson but still had not received his Ph.D. degree when he met Watson. In fact, his discovery of the structure of DNA was part of his thesis work! Crick's background was in physics; he knew little about biology, but was an eager and quick learner. His progress after 1953 was so rapid that within a few years he had become the widely acknowledged leader of molecular biological exploration. (He is now on the faculty of the Salk Institute in California.) His incisive analytical sense and articulate synthetic brilliance combined to make him much sought after by colleagues.

Another interesting aspect of the man was his unabashed atheism. He had entered biology from physics, explicitly committed to showing that the mysteries of biology could be solved using the principles of chemistry and physics. This would allow him to confound the vitalists. Vitalism was the ageless espousal of the existence of inexplicable, mystic vital forces animating living creatures. Crick, with Watson, was now on the threshold of a discovery that would show that the explanation of all inheritance resided in a relatively simple chemical substance. Perhaps never before had one so firmly committed to striking a blow for rationalism been so resoundingly successful — in the eyes of the rational. The vitalists, of course, live on untouched.

Watson and Crick's road to discovery is too technically complex for a detailed narrative here. Instead, I am going to recount the principal events. The action took place mostly in a room at the Cavendish, a focal point for banter, laughter, and the digestion and manipulation of data and ideas.

October 1951. Crick and Watson meet and begin to discuss DNA.

November 1951. Rosalind Franklin obtains x-ray diffraction data showing that DNA consists of two chains running in opposite directions, but does not recognize this. (This was the critical evidence finally seen by Crick a year and a half later, in February 1953, and correctly construed.)

November 1951. Crick and Watson begin to build a model based on inadequate x-ray data. Wilkins and Franklin visit them at Cambridge, and their criticism demolishes the model. Crick and Watson are asked to desist from further work on DNA by Sir Lawrence Bragg, the laboratory head. He does not wish to step on the toes of the King's College group. Watson goes back to working on phage and visiting in Europe. Crick returns to his thesis on protein structure.

May 1952. Franklin makes x-ray pictures of DNA that shout helix, but again, she fails to interpret the evidence correctly. At this time, she is the only person working full time on DNA.

Spring 1952. John Griffith, a mathematician and colleague of Crick, produces calculations indicating that, among the four bases, adenine would tend to attract thymine, and guanine would tend to attract cytosine. That is, these particular pairings may reflect a natural affinity. (Griffith, incidentally, was a nephew of the Griffith who, as we saw in Chapter I, discovered pneumococcal transformation.)

May 1952. Chargaff visits Cambridge, meets Watson and Crick, is not impressed. However, he tells them of his measurements of nucleotides in DNA, showing that the amount of adenine always equals the amount of thymine and the amount of guanine always equals the amount of cytosine. Watson and Crick have obviously not taken these findings seriously enough. Now Crick does. He's suddenly *forcibly* struck by the meaning of Chargaff's 1:1 nucleotide ratios. Adenine, a two-ringed structure, must always pair with thymine, a one-ringed structure. Guanine, a two-ringed structure, must always pair with cytosine, a one-ringed structure. Pairing is *complementary*, with large always opposite small and small always opposite large. It occurs to Crick that this kind of pairing could be the basis of *duplicating* a molecule! It is enormously gratifying that Chargaff's hard data agree with Griffith's imaginative guessing about base affinities.

Crick, in talking to Chargaff, cannot remember which nucleotides are expected to pair with which in Griffith's scheme. Later, when he reports to Griffith, he can't remember what Chargaff's pairings are! When this is straightened out, the data agree.

November 1952. Linus Pauling gets to work on DNA, bringing up guns that will prove to be of small caliber.

January 1953. Pauling's model for DNA, in manuscript form, reaches Watson and Crick. To their relief, it contains errors and need not be taken seriously.

Sir Lawrence Bragg now decides to unleash Watson and Crick. Watson is shown some x-ray pictures by Franklin; these convince him that DNA is both double and helical.

February 1953. Crick and Watson resume model building. Crick now sees Franklin's data of November 1951; he immediately perceives that her x-ray photos most likely mean that DNA is a double helix in which the chains run in opposite directions, a very telling piece of evidence. Let's look at this a bit more carefully.

The phosphate-sugar backbone of DNA has a direction in the same sense that a sentence in English has direction.

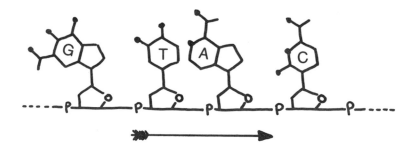

Crick's insight was that if one chain runs in one direction, its mate runs in the opposite direction.

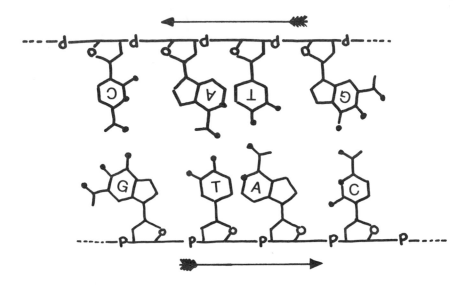

The reason for the chains running in opposite directions will soon become apparent.

February 1953. Jerry Donohue, a former colleague of Pauling, now working with Crick and Watson, tells Watson that the biologically *natural* forms of the bases make adenine fit snugly with thymine only, and guanine with cytosine only. The snug intimacies between these pairs are made by hydrogen bonds. Hydrogen bonds are weak, but enormously important bonds in which pairs of chemical groupings on the nucleotides share a hydrogen atom between them. These hydrogen-atom-sharing groupings on the nucleotides are indicated in my pictures by dots protruding from the base's rings. Note the perfect complementary interaction.

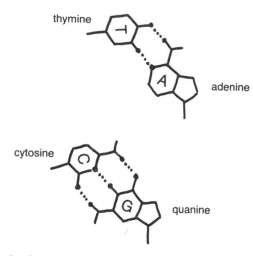

When the hydrogen-bond-linked adenine-thymine pair and the hydrogen-bond-linked guanine-cytosine pair are superimposed, they occupy exactly the same amount of space. And this, you will perceive, fits perfectly with Chargaff's data, which say that *all* of the adenines in DNA occur in the same amount as all the thymines, and all the guanines occur in the same amount as all the cytosines. The theory of the *shape* of the bases and their hydrogen bonding affinities thus merges harmoniously with the facts of base composition of DNA.

On the basis of Crick's interpretation of the x-ray evidence, Crick suggests to Watson that he try pointing the bases *inward*, backbone outward. Although both knew that the bases stuck out at right angles to the sugar-phosphate backbone, it was not clear until near the end whether they stuck inward, toward the central axis of the helix, or outward! They tried the bases inward on the model.

February 28, 1953. Watson makes cardboard cutouts of the four bases, reshaped according to Donohue's suggestions. The two complementary pairs superimpose beautifully. Now when they're attached to a section of backbone of each of two chains forming a helix and are arranged to point inward, they fit perfectly. Crick sees immediately that this way of pairing can only work if the chains run in opposite directions. So the model now demands what the x-ray data demand.

First week of March 1953. Watson and Crick begin to build the final model. The pieces all fit easily. With rising excitement, they are sure they have it. They feverishly complete the model and issue invitations to their colleagues to come admire it — and criticize it.

March 7, 1953. A string of experts visit the Cavendish from other labs and from London. They are compelled by the perfection and beauty of the model to agree that the model builders are right. There's the ring of rightness, the feel of truth; simplicity is, indeed, a sign of truth.

April 1953. Linus Pauling visits. To the delight of all, he expresses good feelings about the structure.

April 25, 1953. The first publication appears.

❖

Here are the two basic features of the three-dimensional DNA:

1. *Four bases are hydrogen-bonded so as to make complementary pairs.* Any base can be attached to either backbone chain so long as its proper complementary mate is opposite it: adenine opposite thymine, cytosine opposite guanine. The width of all pairs is identical, so the chains (the backbones to which the bases are attached) are exactly parallel. The plane of the flat surface of the paired bases is perpendicular to the backbones, making the base pairs like the steps or rungs of a ladder. The bases are, of course, firmly affixed to the sugar of the sugar-phosphate units of the backbone.

2. *Two backbones with opposite chemical directions go up in one chain, down in the other.* This condition is forced on the chains by the rigidly attached bases. To achieve hydrogen-bonded complementary pairing with their opposite partners, as above (and shown on page 110), one complete chain must be reversed 180° so that the two chains form side pieces of the ladder running in opposite directions, as shown on page 109.

You'll also note in the drawing on page 109 that the hydrogen bonds aren't formed between the appropriate base pairs on the chains. That's because one final change in shape is needed to bring them into just the right configuration. Take hold of the top and bottom of the ladder and twist the top counterclockwise, holding the bottom firmly. The ladder slowly becomes a double helix, and the hydrogen-bond-forming dots meet up with each other and form the base pairs! Now the two backbones spiral upward, connected at precisely regular intervals by the paired bases whose planes are perpendicular to the central axis of the double helix.

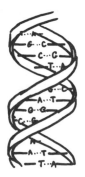

(Note the different way of depicting DNA here. The ribbon-like structure is the backbone, the lettered rods are the bases, and the dots are hydrogen bonds.)

In real life you wouldn't have to force anything. The two chains are flexible and naturally assume hydrogen-bonded helical configuration. This is the natural, most relaxed state of the molecule where all parts are energetically most comfortable. And there it is, the carrier of life's information!

It was quickly obvious to Crick and Watson that if their model was right, and they were sure it was, it was making a clear statement of how it duplicated itself. Hydrogen bonds are very special kinds of bonds. Though strong enough to hold the bases together, they are weak and easily broken. All the other bonds in DNA are strong, permanent, and unbreakable. You can see that breaking the hydrogen bonds between the bases will cause the chains to unwind and separate. Then, nucleotide units (base-sugar-phosphates) are laid down along each strand.

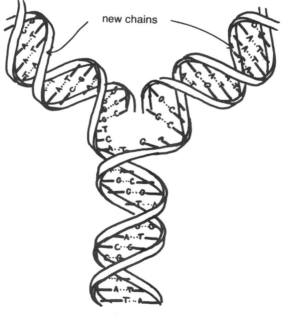

new chains

old DNA

As before, each new base pairs with its correct partner. Once precisely located by base pairing, the sugar-phosphate parts of the nucleotides can be linked to each other to make a steadily extending new backbone. The result is two double helices where there was one, and both are perfect replicas of the original. Here, revealed at last, is a chemical model of one of nature's most astonishing accomplishments, the perfect replication of genetic information for subsequent generations.

Replication, yes. But where is its language? The model is mute about that. The nature of its information, of the instructions for making a cell, still lies hidden. We must search further.

CHAPTER NINE

The Watson-Crick Model Is Confirmed

DNA chains can be separated and will reunite. Matthew Mesel-son and Franklin Stahl demonstrate DNA's mode of replication in living cells. Arthur Kornberg discovers how cells make DNA.

THE WATSON-CRICK MODEL of DNA was only a theory in 1953. The intuitive feeling shared by almost all the experts, that the structure was too profoundly satisfying not to be right, did not, of course, constitute proof. Could a structure so singularly beautiful and utilitarian, and accommodating all important chemical and physical facts, possibly be wrong? Science knows, of course, that it could, and it is the scientist's job to be skeptical. The more important the hypothesis, the greater the effort to prove or disprove it. The Watson-Crick model provoked a burst of experimentation in Europe, America, Japan, and the Soviet Union that continues to this day. Validation of the model came from many different quarters, many ingenious experiments. This chapter will give just three examples.

When dissolved in water, DNA has the physical properties of a long, rigid rod, which it is. If a solution of DNA is heated slowly, the rod-like quality disappears and the DNA behaves like tangled, collapsed single strands. Separating the strands, or chains, of a DNA double helix is called denaturation, or melting. The increasing temperature sufficiently agitates the weak hydrogen bonding between the bases so that they break apart. The separated chains of DNA become randomly tangled coils, their hydrogen-bonding capacity frustrated. The transition from helix to coils can be accurately measured.

What is truly astonishing is that DNA can *re*nature. Those tangled single chains can find their partner chains, pair up their nucleotides in perfect register, and reconstitute a perfect double helix! All that is required is to lower the temperature very slowly, to allow time for hydrogen bonds between complementary pairs of bases to re-form. This kind of physical manipulation of DNA strongly supports the Watson-Crick model's stipulation that two chains are held together by hydrogen bonds.

Denaturation-renaturation has great value in biology. It's useful, for instance, in assessing species relatedness. If DNAs from different species are denatured and the resulting single chains are mixed together, there will be cross-renaturation. "Hybrid" double helices will form frequently with closely related species, but infrequently with distantly related species. The technology involved in this kind of analysis is clever.

First, we make the DNA of an organism of one species (A) radioactive (indicated by *). This is easy, because the cells are constantly making new DNA for growth or for cell replacement. Give them a shot of radioactive phosphorus, and

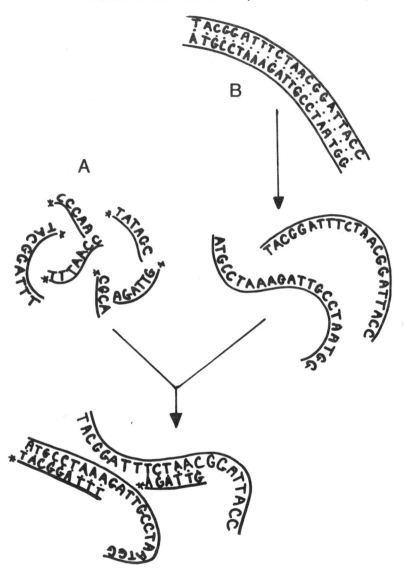

their cells pick it up and build it into their new DNA. Cells
can't tell the difference between radioactive phosphorus and

normal phosphorus. They pick it up the way an absent-minded bricklayer at a building site might unknowingly add an occasional cement brick to the wall he's building — if cement bricks were added randomly to his supply.

Next we extract the radioactively labeled DNA from the cells and break it into shorter lengths by ultrasonic vibrations. We mix it at high denaturing temperature with the nonradioactive, long DNA of another species (B). The temperature is then slowly lowered to encourage renaturation of the two DNAs. The DNA molecules constantly bump into each other, "feeling" for stretches of nucleotides that will form hydrogen-bonded pairs. These regions are called *regions of homology*. Homologous regions, with adenine-thymine and guanine-cytosine pairs properly hydrogen bonded, become stable helices. They steadily accumulate as more and more of them meet. Because long pieces of DNA are easily separated from the short pieces of radioactive DNA originally added, the amount of radioactivity attached to long strands can easily be measured. This provides a direct measure of homology.

Tests of homology between DNAs from different organisms have been a powerful tool for confirming the conclusions of evolutionists based on less precise methods. It is of interest in the light of the brilliant intellectual performances we discuss in these pages that most monkey DNA is 90 percent homologous with human DNA, and chimpanzee DNA is 99 percent homologous with human DNA.

❋

The second experiment that tests the Watson-Crick model is quite different from the first. This one attempts to verify the postulated replication mechanism. The model asserts that when DNA duplicates itself, it does so by conserving its two original strands in the two double helices it makes, thus:

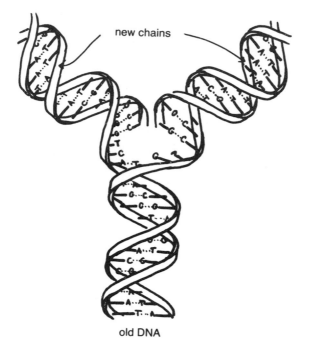

old DNA

This might be provable. Matthew Meselson, then and now a Professor in Harvard University's Department of Biology, and a student, Franklin Stahl, thought they knew how to do it. The trick was to be able to distinguish between *old* DNA and *new* DNA, that is, between parental DNA and new DNA made as replication was occurring. At first glance, that seems ridiculous. DNA is DNA whether its molecules are old or new. But they had an idea. They knew that physicists made atoms heavier by adding neutrons to their nuclei. For instance, nitrogen normally has an atomic weight of 14 (^{14}N) but can be made with an atomic weight of 15 (^{15}N). Natural carbon (^{12}C) can be made ^{13}C. These atoms are measurably heavier, yet the body treats them as though they were the natural, light forms of the atoms (the same way radioactive compounds are handled).

Meselson and Stahl reasoned that if they could get bacteria to consume heavy atoms, they would eventually contain heavy DNA. Bacteria must make DNA before they can make more of themselves, and if the only source of material for making DNA is heavy (they used heavy ammonia ^{15}NH$_3$), their DNA would *have* to be heavy. Once the experimenters obtained a population of bacterial cells all with heavy DNA, they could switch them to fluid containing light atoms (light ammonia, ^{14}NH$_3$) and the bacteria would begin to make new molecules of DNA, now light. Thus, *heavy* DNA would be equivalent to *old* DNA; *light* DNA, to *new* DNA. Thus was *age* of DNA converted to *weight* of DNA for purposes of experimentation.

Life's delicate processes inside cells are not, as we've said, able to detect a gain in weight of their proteins or DNA. We humans, however, have devised exquisitely delicate methods that allow us to distinguish clearly between, and even to separate, heavy and light molecules. You may have guessed, correctly, that this separation is done by gravity — the exaggerated gravity attainable in a centrifuge. When concentrated solutions of certain salts are subjected to very high centrifugal force for days, the salt solution forms a gradient, with the least dense solution at the top of the tube, and a steady increase in density to the bottom of the tube.

Now, if a salt solution is similarly centrifuged to generate the same gradient, but this time DNA is included in the solution, the DNA will seek the portion of the tube where the density of the salt solution corresponds to its own density. After the prolonged centrifugation, it will appear as a thin disk (or, if viewed from the side, as a band) near the middle of the tube.

Meselson and Stahl were ready. They had their bacteria grown in heavy ammonia ($^{15}NH_3$). They now transferred them to a "light" medium (ordinary $^{14}NH_3$). Then, after exactly the time it took for the cells in the population to divide once, they withdrew a sample of bacteria. At the time the second division should occur, they took a second sample. From each sample, they extracted the DNA, mixed it with salt solution, and centrifuged this mixture for three days.

Here's what they found. Just before transfer of the heavy bacteria to normal (light) nitrogen, there was only one band of DNA, all heavy, as one would expect.

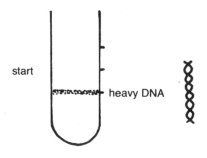

start

heavy DNA

After one average generation of DNA-making in light nitrogen, *all* the DNA had shifted to a lighter band that was intermediate between where all heavy DNA would appear and where all light DNA would appear.

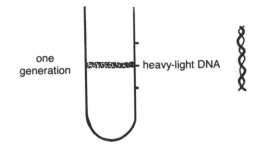

This meant that the starting heavy chains had all separated and had laid down new light chains along themselves. That explained why they were of exactly intermediate weight: half light, half heavy. After *one more* generation (another doubling of the population and another doubling of DNA), half of the DNA was at the half-heavy, half-light density, and half was all light.

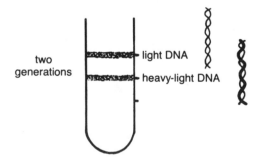

This generation started with the half-heavy, half-light DNA made in the first generation. The heavy chain now separated from the light, and new light nucleotides were laid down along each. This produced two double chains; one contained one light chain and one heavy chain, and the other contained two light chains.

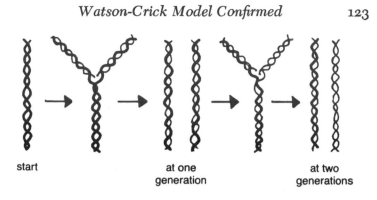

start   at one   at two
generation   generations

So it was that an imaginative application of physics that permitted a direct observation of the DNA-copying process in living cells confirmed the predicted mode of replication implicit in the Watson-Crick model.

❊

One of the most impressive contributions made by biochemists to the knowledge of the gene came from the laboratory of Arthur Kornberg, Professor of Biochemistry at Stanford University Medical School in the early 1950s. His discoveries furnish the third example of early support for the Watson-Crick model. Kornberg has devoted much of his career to the study of how DNA is made in cells. His methods were classic: he broke cells open with as little damage as possible, added some radioactive nucleotides to the cell innards dispersed in fluid, incubated the mixture for a time at body temperature, stopped whatever might be happening with a chemical that killed enzymes but didn't hurt DNA, and then looked to see whether any radioactivity had got into DNA. A positive result, it was hoped, meant that a small amount of DNA had been made using the radioactive nucleotides as building blocks.

Kornberg did find that small amounts of radioactive nucleotides found their way into DNA. The next step was to fractionate the crude cell innards, that is, separate them into different parts. This was done using a variety of known procedures. Each fraction obtained was tested for its ability to make DNA, by adding radioactive nucleotides and, as before, measuring the amount of radioactive DNA after an incubation. If one particular fraction incorporated radioactive nucleotide into DNA more actively than others, that fraction was further fractionated or subdivided, and so on. With this method, Kornberg and his associates relentlessly tracked down a fraction that made DNA vigorously and seemed to be able to do nothing else. He had obtained the enzymes that make DNA. Thus, Kornberg focused on the cell's DNA-making machinery in very much the same way that Avery had pursued the transforming material ultimately shown to be DNA.

Kornberg won the Nobel Prize in medicine for his work on DNA synthesis in 1959. The enzyme he discovered is called *DNA polymerase.*

The enzyme, whether working for the scientist making new DNA in the test tube or operating in the living cell, uses the nucleotides as building blocks only when they are supplied in a very special form.

Look back for a moment and recall (Chapters I and VIII) that the units of DNA, the nucleotides, are made up of a base rigidly attached to a sugar, in turn attached to a phosphate. The single DNA chain is a regularly repeating chain of nucleotides: base-sugar-phosphate, base-sugar-phospate, and so on.

But to make DNA, you can't put together chains of base-sugar-phosphate for free. You need energy. It's like using

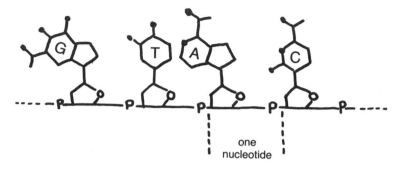

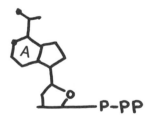

bricks to build a house: the bricks will not spontaneously build themselves into a house; they need the energy supplied by the bricklayer. Each nucleotide building unit is supplied with an energy-giving handle attached to it, and as the unit falls into place in the growing chain using the energy of the handle, the handle is discarded. The energizing handle is a double phosphate, called pyrophosphate (PP), attached to the familiar phosphate of the nucleotide.

Such nucleotides with two additional phosphates attached are called triphosphates: adenosine triphosphate, guanosine triphosphate, thymidine triphosphate, and cytidine triphosphate. (If you think these names are difficult, just think how considerate nature is to have written the language of life in only four letters.) Each time a triphosphate meets up with the enzyme DNA polymerase, the enzyme captures it, carries it to the lengthening end of the DNA chain, inserts its base-sugar-phosphate onto the end of the chain, and then discards the double phosphate as waste once its energy has been used to add the new link to the chain.

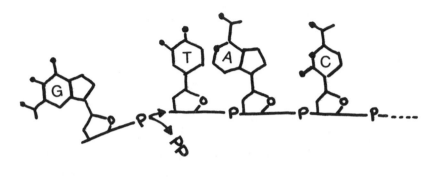

*

Kornberg's enzyme needed one additional essential ingredient, along with the four triphosphates, to be able to make DNA. As the Watson-Crick model clearly demanded, it needed a template — some DNA to copy!

Without DNA to copy, the enzyme would be stringing beads meaninglessly, lining up nucleotides like a mad author writing gibberish. The enzyme alone is uninformed; it lacks the knowledge to turn out biologically meaningful DNA. The enzyme must have real, information-rich DNA to *copy.* Give it some good double-stranded DNA and a plentiful supply of the four triphosphates, and sure enough, the helix unwinds, the hydrogen bonds break, the two chains separate, the enzyme brings up the triphosphates, their bases pair up with those on the added DNA, and new chains lengthen along the separated chains. When the job is complete there are two double chains where there had been one.

With an enzyme, energy, and a template, new DNA was

created in the test tube exactly according to the expectation of the Watson-Crick model. The system copied any added DNA in its own image. The template had complete control of the type of DNA made. Kornberg could add all sorts of odd amounts of the four triphosphates, or odd nucleotides not known to enter DNA, and the system paid no attention; it proceeded to copy the template perfectly, rejecting the irrelevant material.

*

Three groups of experiments, then — those of Meselson on the way DNA replicates in the cell, those of Kornberg on DNA replication *outside* the cell, and those on the denaturation and renaturation of DNA — all combined to provide formidable proof for the Watson-Crick model. As a variety of other evidence accumulated, the Watson-Crick model became the central reality in the advance of biology.

CHAPTER TEN

Genetics Dissects the Gene

Seymour Benzer redefines the gene by complementation test and finds gene is divisible into many parts. Colinear relationship of gene and protein proved by Charles Yanofsky using missense mutations and by Sydney Brenner using nonsense mutations.

AT THE TIME of the appearance of the DNA model in 1953, the gene still lurked in shadows. The linear layout of genes as determined by recombination frequency and confirmed by interrupted mating, and the linear molecular structure of DNA suggested that information in genes, and gene sequence as well, had their physical basis in the sequence of nucleotides in DNA. Dimly discernible, too, was a structural parallelism between DNA and protein. As we noted in Chapter V, the sequence hypothesis was becoming, in the early 1950s, the guiding theory in the field. This hypothesis stated that there must be some simple linear relationship between the nucleotides in DNA and the amino acids in protein.

Classic genetics applied to *Drosophila,* mice, peas, or other complex test subjects was limited in the precision of its discrimination, just as an ordinary light microscope can magnify so much and no more. This approach had determined the relative linear position of genes on the chromosome and had roughly estimated distances between them. It was believed that recombination, whereby segments on one chromosome (one DNA double helix) were exchanged (by breakage, crossing over, and reunion) with segments of another contiguous chromosome, only occurred *between* genes and not within single genes. A new kind of genetics with a resolving power comparable to that of the electron microscope was now poised to destroy that illusion.

The name of the new genetics was *fine-structure genetics*. It was really no more than the stretching of bacterial and viral genetics to their logical ends by using increasingly refined techniques. The enormously enhanced power of the new techniques was based simply on their ability to observe rare events. Phages and bacteria came in tremendously larger numbers than peas, fruit flies, or mice, and selection techniques allowed the detection of a few mutants, or a few normal recombinants, among many millions of cells. That made all the difference. Remember that in a cross between two organisms carrying mutations in different locations on their chromosomes, the nearer those locations are the *less* frequently will a recombination occur between them. Yet given a large enough population and a means of detecting extremely rare recombinants, the chance of finding recombinations between extremely close mutations in different genes remains good.

Seymour Benzer, a Professor at Purdue University in the early 1950s, single-handedly developed the technology of fine-structure genetics, quietly applied it, and astonished his colleagues around the world with a wholly new vision of the gene.

His prodigious industry and experimental virtuosity

proved that recombination between chromosome pairs could occur *anywhere* along their lengths. He killed the idea that recombination only occurred *between* genes, as though the gene itself resisted recombination within its boundaries.

Benzer has an amused, sleepy, self-deprecating manner, seasoned lightly with skepticism and worry. He was trained as a physicist and is now experimenting in neurobiology at the Salk Institute in California. His fierce curiosity and unorthodox approach together with a facile wit give him a special place among his colleagues.

The occurrence of recombination at any point along the chromosome meant that the gene had to be redefined. The old definition of gene was based on the assumption that nothing less than a gene could recombine; that the gene itself was the minimal unit of recombination; that the gene, in fact, was indivisible. In the analogy of the freight train in Chapter V, the box cars (genes) had always been assumed to be indivisible, with switching possible only at the couplings between them. Now, *sections* of box cars were being switched across the tracks. So died the indivisible gene, and Benzer had to find a new way to define it.

Let's examine Benzer's problem of reassessing the gene by looking at a specific problem. If two mutants are defective in two different functions, then they are defective in two different genes. In such a case, there's no problem in assigning one function to one gene, and the other function to the other gene. There's no ambiguity. The genes are defined in terms of their specific functions. But suppose you have two mutants defective in the *same* function. *Are they necessarily defective in the same gene?* Well, they may be or they may not be, and that's the nub of the problem. Take the case of two mutants, both unable to make the amino acid valine. (They both need valine to grow.) Valine is normally made by a series of steps, as we learned in Chapter V; each step requires a different enzyme, and each enzyme is controlled by a different gene. The mutants could both be damaged in the *same* enzyme, or each could be damaged in a *different* enzyme. These two possibilities point up the ambiguity created by Benzer's demonstration of recombination within genes.

To understand this difficult genetic situation, let's consider two small, identical shoe factories. Each factory has two workers, A and B; A makes uppers and B makes soles.

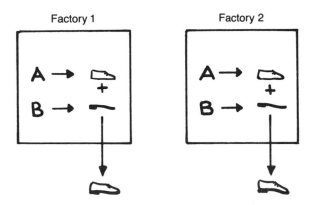

Now suppose that after a flu epidemic, both factories stop producing shoes. Is the stoppage due to indisposition of the *same kind of worker,* that is, do both As or both Bs account for the shoe shortage? Or is the stoppage due to disability of *different workers,* that is, A in one factory and B in the other?

Here's Benzer's solution, applied to shoe factories. Put

the personnel of one factory *into* the other factory and see whether finished shoes begin to appear. If As had been disabled in both factories, uppers could not be made so no shoes would be produced. If, on the other hand, *different* workers were incapacitated in the two factories, the combined operation would have one functional uppers-maker and one functional soles-maker and shoes would still be made.

Now let's see how the solution works in practice with genes in cells instead of workers in factories. Return to the case of two mutants, both unable to make valine. We wish to know whether they are damaged in the same gene or in different genes. The critical test is to put the relevant genes together into *one* cell. This must be done in such a way that the genes are stably maintained, allowing us to measure the performance, or function, of the cell. Such a diploid state (two sets of genes in one cell) can be achieved by the techniques of microbial genetics. If the bacteria still fail to make valine, we may conclude that the mutations had occurred in the same gene in both mutants. (Both As or both Bs in the shoe factories are incapacitated.)

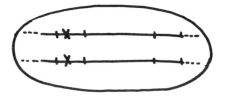

If, on the other hand, the bacteria can now make valine, then the mutations must be in different genes, and the combined genes each provide one good gene. (A is incapacitated in one factory, and B is incapacitated in the other.)

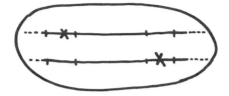

There must be one good copy of every gene for an organism to perform all its functions normally.

This test for defining the gene is called the *complementation* test, because two different undamaged genes complement each other and restore function when they're together in one organism.

This new way of assessing the meaning of gene is designed, for the purposes of experimental genetics, to relate more precisely the changes in genes to functional alterations in an organism.

Benzer's work was done with phages, but his discoveries are equally applicable to bacteria, from which I have chosen my illustration. Seeking out increasingly rare recombination events, Benzer finally concluded that, within a given gene, many recombinations could occur. His calculations indicated that some recombinations occurred so infrequently that they must be extremely close to each other in a gene. In fact, if the gene were DNA, it looked as if recombination might be occurring between adjacent, or near adjacent, nucleotides. Put another way, in a gene 1,000 nucleotides long, it looked as if 1,000 different recombinations could occur. Fine-structure genetics, indeed!

The essence of Benzer's accomplishment may be likened to the detection of increasingly frequent errors in a pair of identical paragraphs. If we think of a paragraph as equivalent to a section of a chromosome, the sentences therein are equivalent to genes. Before Benzer, sentences were, like our box cars, thought of as inviolable. One sentence could recombine with another sentence in identical paragraphs, by breakage and reunion *at the periods.* Benzer showed that breakage and reunion could occur anywhere *within* the sentences, indeed *between any of the letters.*

Genetic analysis could now provide an unequivocal definition of a gene, and could spot the exact location of mutations, even when very close together. It ought, then, to be able to approach a proof of the sequence hypothesis: the idea that there is a direct linear relationship between nucleotides in DNA and amino acids in protein. Such a proof required the demonstration of a simple, direct *linear* relationship between the location of single nucleotide mutations in a gene and the location of amino acid changes in the protein made from that gene. Fine-structure genetics and the new methods of breaking protein chains and determining their amino acid sequence now made this proof possible.

The operative word was *colinearity.* One needed to show that a gene's nucleotide sequence was *colinear* with the amino acid sequence in the protein specified by the gene.

By the mid-1950s, the best minds at the front of molecular genetics were hotly competing to be the first to prove colinearity. Most notable were Sydney Brenner at the Cavendish Laboratory in Cambridge, and Charles Yanofsky at Stanford. Yanofsky won the race, but Brenner finished soon after with such an elegant proof that the meaning of winning was blurred.

Yanofsky, a tall, taciturn, modest scientist, focused on an enzyme in the bacterium *E. coli* that is needed for the synthesis of the amino acid tryptophan. (This enzyme, tryptophan synthetase, is really two proteins linked together, but he concentrated on just one of the two.) Based on extensive recombination studies, Yanofsky calculated that the gene was about 1,000 nucleotides long. He also knew a lot about the enzyme itself. It was 280 amino acids long, and he was rapidly identifying every amino acid in the chain. It was a prodigious task.

If the bases in DNA code for, or specify, the amino acids in protein in a simple way, then with a gene 1,000 bases long and a protein 280 amino acids long, 3 to 4 bases must specify 1 amino acid (1,000 divided by 280).

Francis Crick and others had for some time favored three bases for 1 amino acid on purely theoretical grounds. The argument was as follows. If genes are made of DNA, and DNA is a long chain, and if there is some simple linear relationship between DNA and the protein it specifies, then some fixed number of nucleotides in a DNA chain must determine which of the 20 amino acids will be located in a particular position in the protein chain. There are four nucleotides in DNA and 20 amino acids in protein. Obviously, four nucleotides cannot unambiguously code for 20 amino acids. Can four nucleotides taken in pairs do it? No, that's still not enough, because there are only 16 possible unique combinations of four nucleotides taken two at a time. What about three nucleotides? Now we have enough. There are 64 possible unique combinations of four nucleotides taken in groups of three. This is well in excess of the 20 required, and this is why theory favored three nucleotides per amino acid.

Thus, quite early in the experimental phase of applying fine-structure genetics and protein chemistry to the problem of the *molecular* relationship of DNA to protein, theory and practice were in good agreement.

Yanofsky now induced mutations in the tryptophan synthetase gene, producing a series of mutants with function-less enzymes. He used the complementation test to prove that all mutations had occurred in the same gene. The mutations were all point mutations, that is, single nucleotide changes in the DNA. It is important to emphasize that the mutant bacteria did produce enzymes, even though they were damaged and nonfunctional, because it was essential that Yanofsky be able to find and purify the enzymes so that he could break them down and work out their amino acid sequence. This difficult step was needed to know the normal amino acid sequence and the changes that occurred in the mutants.

Yanofsky found that the enzyme from each mutant had an amino acid change, and that the change occurred at a different location in each mutant. Yanofsky then carefully analyzed recombination frequencies to map the location in the gene of each mutation in each mutant.

Now he was ready for the coup de grâce. He laid out a map of the gene, with the mutations marked; alongside this, he placed the sequence of the 280 amino acids of the enzymes, with the amino acid substitutions of each mutant marked. He looked to see whether there was a linear correspondence. We wouldn't be looking over Yanofsky's shoulder at this moment if he hadn't hit the jackpot! In each case, the genetically mapped mutation corresponded linearly to the location of the amino acid substitution in the protein. This is a rough sketch of Yanofsky's beautiful results:

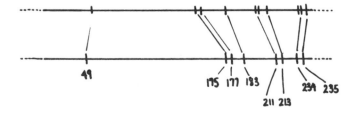

The upper line is the relative location of ten mutants within the tryptophan synthetase gene, determined by Yanofsky using fine-structure mapping. The lower line is the physical location of the mutants' amino acid substitutions in the tryptophan synthetase protein itself. The numbers are the locations of the amino acids, starting at one end of the protein chain and counting through to the last, number 280. A clean, simple, unequivocal plunge into the cell's depths to retrieve a bright truth: the proof of colinearity. It is hard in this account to convey what a colossal technical job was done by Yanofsky; we have described years of grueling work in a page. For a decade, workers in the field had been acting as though the sequence hypothesis was fact. They knew what Yanofsky and others were after, so the truth had the look of a friend.

*

The mind of Sydney Brenner illuminated much of molecular biology during its liveliest period. The loose professional companionship of Brenner and Crick at the Cavendish generated some of the best ideas and some of the cleverest ways to test them.

By 1957, Crick had become the leader of the new biology. His mailbox at the Cavendish was filled with letters and manuscripts from scientists in Europe, the Soviet Union, the United States, and Japan who sought his critical attention. Crick and Brenner were reciprocal sounding boards, and their witty and fertile conversation dominated lunch time, tea time, and much of the day in between. Sydney Brenner was a refugee from South Africa, short, tough, aggressive, voluble, and flamboyant.

He approached the colinearity problem using a system entirely different from Yanofsky's. The kind of mutation

Yanofsky used in the tryptophan synthetase system of *E. coli* is called *missense*. A missense mutation is one in which a different amino acid is substituted for the one normally in place in the protein. The complete protein, although defective, is still produced. Brenner, however, used *nonsense* mutations, which his work helped to characterize. A nonsense mutation is a change in DNA that the protein-making machinery cannot cope with, cannot read as amino acid.

We shall learn later that a protein is made by attaching amino acids one at a time to a growing chain, starting at one end of the chain, very much like stringing beads. DNA supplies the sequence instructions. When a nonsense mutation occurs in DNA, the chain-making operation stops. As a result, the partially completed chain is cast off. The synthesizing machinery starts again and again, trying to make a complete chain and throwing off the unfinished product. A nonsense mutation is an incomprehensible instruction in a sequential process. Imagine a factory of workers stringing necklaces. Each worker (a protein synthesis machine) is supplied with a set of instructions (DNA) specifying an exact sequence of colored beads. Suppose that one worker gets instructions that neglect to specify the color of a certain bead. The worker, irritated, throws the unfinished necklace aside and starts a new one. (In the same image, a missense mutation is an instruction in which a bead of a different color is unexpectedly specified for the usual color. Because the worker has a supply of beads of all colors, he simply strings the anomalous bead and proceeds to finish the necklace.)

Brenner had long favored phages as experimental subjects and had been using several of their proteins to try to show correspondence between mutation and amino acid sequence, as Yanofsky was doing. The protein that proved to be most interesting was the "head" protein. This is the protein that makes up the principal part of the virus inside which the DNA is packed.

There's a lot of this protein, and when nonsense mutations are induced in the phage chromosome and mutants are isolated, a number of them have defective head proteins. Under the microscope one sees a lot of heads in various stages of completion, like the parts of a prefabricated house not yet put together.

It occurred to Brenner that the nonsense mutations were stopping the completion of head protein at different points along the protein chain. If he knew the linear location of the mutations in the DNA, he might be able to discover whether they corresponded to the *lengths* of the partially finished proteins. This would be equivalent to lining up missense mutations with the points where amino acids were changed in protein. Brenner mapped his mutants.

Mapping in phages, incidentally, is done by recombination, as in other organisms. To "mate" two phages to get recombination between their DNAs, you infect bacteria simultaneously with two mutant phages. While the phage DNA is still naked inside the bacteria, recombination between phage chromosomes occurs very much as we've de-

scribed it in mating bacteria. Normal recombinants may be
detected at extremely low frequency. This system was used
by Seymour Benzer in establishing fine-structure genetics.
Brenner followed this procedure to map his nonsense mutants. He also developed ways of determining the length
of the incomplete head proteins. His results were as beautiful as the experimental conception: the lengths of the unfinished proteins in each of the mutants corresponded to
the mapped locations along the gene of the various nonsense mutations.

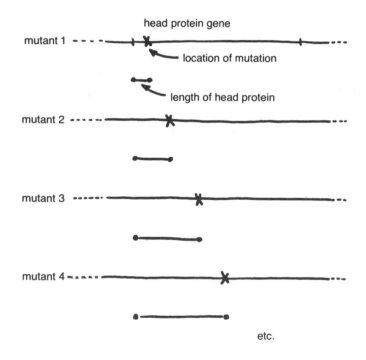

Yanofsky's and Brenner's brilliant proofs of colinearity
using the ultimate refinements of genetics and protein

chemistry began in the mid-1950s and finally succeeded in the early 1960s. The technology upon which their work depended — sequence analysis of the amino acids in protein and fine-structure genetics — was slow in maturing and technically very difficult.

❋

I experienced one facet of the spirit of the colinearity struggle in Cambridge at the Cavendish Laboratory in the spring of 1958, when the contest was at fever pitch. An American colleague and friend had been writing regularly to Brenner, telling of his steady advances on the colinearity question, implying that success was imminent. Brenner, in replying, was tempted to exaggerate the extent of his progress but forbore. One day a letter arrived from his American competitor saying he'd soon be visiting Cambridge to discuss experiments. Brenner forthwith buried himself deeper in his researches. When the American arrived, Brenner presented an informal stand-up lecture in the lab to which he had invited all the resident staff. Francis Crick, John Kendrew, Max Perutz, Vernon Ingram, Seymour Benzer, George Streisinger, and Paul Doty were all present — all men of outstanding accomplishments who were well versed in the problems in the field. During the course of his carefully prepared lecture, Brenner moved inexorably and eloquently toward incontrovertible proof of the colinear relationship of gene and protein, using numerous slides and demonstrations. Brenner's colleagues were astonished and delighted with his success and felt increasingly sorry for the visitor, who visibly wilted during the presentation.

It dawned first on Crick that the whole presentation was concocted out of whole cloth! Unable to conceal his amusement and embarrassment, he left the room. One by one, the other staff members caught on and similarly departed. Only when Brenner and his disheartened guest were alone was the guest informed that he'd been the victim of a practical joke. I'm sure that never in the history of science had such an illustrious array of scientists been present at the perpetration of a scientific hoax!

＊

Before Yanofsky and Brenner claimed victory, other scientists pursuing biochemical routes had discovered the cell's machinery for making protein. The analysis of that machinery cleared up the mystery surrounding gene-protein relationships, including the principle of colinearity. These developments will be considered next.

CHAPTER ELEVEN

The Protein-Making Machinery Uncovered

Paul Zamecnik's laboratory probes cells for evidence of protein assembly. Enter ribonucleic acid (RNA), as ribosomes, the site of synthesis. Energizing the synthetic process. Devices for insuring the proper (genetically determined) order of amino acids: transfer RNA and messenger RNA.

THE MAIN PURPOSE of the new biology was to reveal the broad principles governing information processing, which meant deciphering codes and reading messages. The traditional biochemists were intrigued by the *mechanics* of these processes. By tradition, too, the biochemists were comfortable working with animals as experimental systems, a practice somewhat distasteful to those who had been seduced by the shiny new microbial systems. Despite these divergent scientific approaches, the biochemists in the course of a few years uncovered the essential features of the protein-synthesizing machinery and, using that machinery, unlocked the final secret of DNA's structure: how it stored and released encoded information. It was a remarkable achievement,

suggesting that in struggling against the unknown, you
never know what weapons will serve you best.

*

It might be instructive at this point to follow the events of
those exciting years through the eyes of a scientist who was
intimately involved in the search. The one that I know best
is me. In 1953 I had just finished a year in the laboratory of
Fritz Lipmann at the Massachusetts General Hospital in
Boston. Lipmann was an expert in the field of energy generation and utilization by cells and, in fact, had received
the Nobel Prize just the year before. I was moving on to a
neighboring laboratory at the same hospital, the Huntington
Laboratories, where a group headed by Paul Zamecnik was
studying protein synthesis. Zamecnik was a contagiously
enthusiastic scientist who had an uncanny sense of the right
paths to follow in experimentation. He gathered around him
from the late 1940s to the early 1960s an exceptionally able
group of young associates who made the Huntington Labs
the world center for protein synthesis.

Zamecnik, who was trained as a physician, believed that
an understanding of protein synthesis was the key to the
long-range problem of understanding cell growth and cancer, the problem to which the Huntington was committed.

If you want to study protein construction in an animal,
it is not obvious how to go about it. An animal is a big mass
of proteins, and it is impossible to know how much of all
the protein is newly made, and how much was there before
you came along. But proteins are chains of amino acids, and
a new protein can't be made without amino acids. That
means you can measure protein synthesis by measuring the
rate of incorporation of amino acids into protein. The key
is having a way to *identify* the amino acid after it gets into

protein. It must be tagged or labeled or somehow given a special identity. That can be done with radioactivity. If an amino acid is made with radioactive carbon (^{14}C) instead of normal carbon (^{12}C), then you can follow it to the ends of the earth.

The use of radioactively labeled isotopes or density-labeled isotopes (such as the ^{15}N used by Meselson, Chapter IX) in normal body constituents is probably the most generally valuable tool biochemists have for probing cell mechanisms. Isotopes were a direct result of World War II's atomic bomb project and related research, and they were made available to researchers in general soon after the war's end. Indeed, a bright young organic chemist named Robert Loftfield made the first amino acids labeled with radioactive carbon (^{14}C) in Zamecnik's lab during this period, giving the group a lead in the exploration of protein synthesis.

Zamecnik's basic method was to inject radioactively labeled amino acids into rats; wait a while to let the amino acids pass from the circulation into the tissues; then kill the animals, remove a convenient organ like the liver, grind it up, separate parts of cells by a variety of means, extract *proteins* from these parts, and measure the radioactivity of the protein in a Geiger counter. If the proteins had become radioactive, that is, if they had incorporated radioactive amino acids into their chains, this was presumptive evidence that the cell's machinery was making new protein chains. (You will note the similarity to the methods of Kornberg in studying DNA synthesis, Chapter IX.)

Zamecnik found that incorporation did occur under these conditions, but it wasn't easy at first to prove convincingly that incorporation meant real synthesis. Radioactive amino acids might, critics said, get stuck to proteins already made, and not be linked integrally in the chain. Much effort was spent proving that real synthesis was being measured by breaking down the radioactive proteins and showing that all the radioactivity could be recovered as amino acids properly linked in sequence in the chains. Once the skeptics had been subdued on that point, the system was poised for exploitation by a steadily increasing number of labs in America and Europe. At the Huntington, we made steady progress in dissecting the system for a decade in a setting characterized by warmth, good fellowship, humor, and intense enthusiasm.

*

A research aid that runs a close second to isotopes is an *ultracentrifuge*. The modern biological ultracentrifuge is a device that applies high centrifugal forces (up to 300,000 times the force of gravity) to mixtures of biological materials — particularly crude mixtures of the parts of cells — for the purpose of separating them and thus purifying them. Meselson and Stahl used such an ultracentrifuge to separate heavy and light DNA in their experiment described in Chapter IX. The ultracentrifuge made its most notable early contribution to the present studies by showing that as liver cells incorporate radioactive amino acids into their proteins, the place where new protein is first found is the *ribosomes*. These little bodies had already been seen in cells through the electron microscope. They are sprinkled throughout the cell's cytoplasm (the space outside the nucleus) like pepper. Ribosomes, it was soon established, are the cell's assembly plants for protein chains. As time goes on, radioactive proteins made on ribosomes drop off and then appear in other parts of cells — delivered there after completion on the ribosome.

When it was learned that ribosomes were the assembly site for proteins, many laboratories began to study their structure. They were found to be extremely complex, consisting of three kinds of ribonucleic acid (RNA) and many different proteins — proteins that are part of their make-up, not the ones in the process of being assembled. But we need not be concerned with ribosome structure. We shall simply look at them for what they do: they are reading machines. They receive information from DNA, which is located far away in the cell's nucleus, and convert it into protein. The ribosomes "translate" the four-nucleotide language of DNA into the 20-amino-acid language of protein.

Ribonucleic acid will figure increasingly in our story, and we should pause here to describe it. It is very similar in basic structure to DNA: four bases are attached to sugar-phosphates to make nucleotides. Three of the nucleotides are essentially the same as those found in DNA: the nucleotides of the bases adenine, guanine, and cytosine. The fourth, corresponding to DNA's thymine, is the nucleotide of the base uracil. Pairing and hydrogen bonding of all the nucleotides of RNA are exactly the same as in DNA. Thus, single RNA chains can pair with single complementary DNA chains and, as we'll learn later, single RNA chains are made naturally as complementary copies of single DNA chains. The sugar of the sugar-phosphates of RNA is different, however. It is *ribose* instead of DNA's deoxyribose. This means that RNA's backbone is chemically slightly different from DNA's, but this difference is not important to our story.

Like the making of DNA chains, the linking together of amino acids to make proteins is a process that must be driven by energy. How this might be accomplished was a

question of intense interest to both Lipmann's and Zamecnik's groups in the early 1950s. When I joined Zamecnik in
late 1953, I went to work on the problem. The techniques
for detecting energy-using reactions in cells that I'd learned
in the Lipmann lab were of particular value in attacking the
problem of energizing protein synthesis. Furthermore,
Zamecnik had found by that time that ATP (adenosine triphosphate, by then recognized as the universal energy-
storing and releasing molecule) was needed for protein
synthesis.

It was expected by most of us working in the field, on
the basis of Fritz Lipmann's researches, that ATP and an
amino acid would somehow meet and react in such a way
that ATP's energy would be transferred to the amino acid.
How that might be done was unknown. In 1954–55 I discovered the answer. Here's how it works. First, recall the
triphosphates: molecules composed of a base, a sugar, and
a phosphate, just like the basic units of DNA and RNA, but
with two more phosphates attached: base-sugar-phosphate-
phosphate-phosphate. These, remember, are the energized
nucleotides that participate in DNA synthesis (Chapter
IX). Now, one of these, ATP, is the cell's general energy
supplier. It does the job just as it does when being built

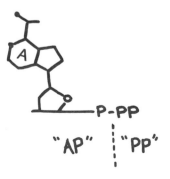

into DNA. That bond between adenine-sugar-phosphate (AP) and the attached double phosphate (PP), when broken, releases a lot of energy as heat. If, instead of breaking the bond and releasing heat, the AP half of the molecule is attached to *another* molecule like an amino acid (aa), that other molecule becomes *energized* or *activated*.

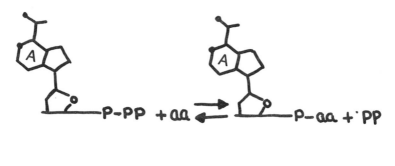

$$AP-PP \ + \ aa \ \rightleftarrows \ AP-aa \ + \ PP$$

This means that the energy that was in the bond between AP and PP is preserved by virtue of AP's attachment to an alternative molecule. Amino acid simply replaces PP and so shares the energy resident in the molecule.

You'll notice in the formula shown in the diagram that I've indicated that the chemical equation is reversible: the reaction can go in either direction, depending on how much of each of the molecules is present. If there's a lot of AP-PP and aa around, it will go to the right; if there's relatively more of the activated amino acid (AP-aa) and PP, it will go to the left. This reversibility was what allowed me to discover the reaction. By adding radioactive PP to material from inside cells, I found that the PP was rapidly incorporated into ATP.

$$AP-\ddot{P}\ddot{P} + aa \rightleftharpoons AP-aa + \ddot{P}\ddot{P}$$
$$\uparrow$$
$$\ddot{P}\ddot{P}$$

Thus, the reverse reaction was used to detect the forward — biologically more significant — reaction.

When we say an amino acid is activated, we mean it has been put into a reactive state, a state of readiness to form a bond with another amino acid. Cells can't accomplish these operations, or for that matter any chemical reactions, without enzymes. Enzymes simply make the events they promote go fast enough to accomplish the cell's purposes. When I say I found that material from inside cells made the activation reaction go, I mean that in that crude mixture of cell materials, enzymes were present whose specific purpose was to perform these functions. These enzymes have been isolated in many laboratories, and it has been found that each amino acid is activated by a separate enzyme!

The activation or energizing of amino acids is the essential first step in protein synthesis. The activated amino acids, as AP-aa's, remain attached to their respective enzymes ready for the assembly of chains that occur on the ribosome. But one very critical step needs to be interposed: a step that will insure the correct, genetically determined, *order* of amino acids in the chains. This is the critical *translation* step, in which the information in DNA, presumably in its sequence of nucleotides, must be translated into an exactly predetermined order of amino acids in protein. How?

In 1953, there were no hints. As knowledge of protein

synthesis grew and the information-carrying role of DNA was accepted, we at the Huntington and others working in the field became convinced that a template for the assembly of protein chains must reside in ribosomes. The template would contain sequence information derived from DNA. This template could not be DNA itself, because DNA was a permanent resident of the nucleus, and protein was made in the cytoplasm.

RNA seemed a logical candidate for template. There's a lot of it in ribosomes, and it is very similar to DNA in chemistry and structure. *If* it could somehow have been born of DNA so as to reflect DNA sequence information, and *if* it could have got to the ribosome to program it, we'd have a plausible scheme. Two very big if's!

It was time for one of science's helpful surprises, and it came in 1956. Of course, a surprise can be relative. This one was a surprise for me and Zamecnik and our colleagues at the Huntington, but not so much of a surprise for Francis Crick.

Let me take it chronologically. Crick had given much thought to the problems of templates and sequence determination. He was troubled by how an amino acid, which had no chemical resemblance whatsoever to DNA, or to RNA for that matter, could nevertheless *recognize* its proper location in a preset sequence. He knew the problem had to have a chemical explanation. He assumed, as we were all forced to at the time, that the resident RNA in ribosomes was the template; there was no other RNA around. Then he made a jump: if amino acids themselves can't recognize an RNA template, why not attach something to them first that *could* recognize an RNA template? RNA, like its sister DNA, can recognize another strand of itself by complementary hydrogen-bonded pairing of nucleotides.

How about amino acids being linked to a short piece of RNA that could then recognize a short length of nucleotides on the RNA template on the ribosome? There was an intriguing new idea. Crick called it the "adaptor hypothesis," in the sense that amino acids were seen as adapted to a template by the fragment of RNA they had become attached to. The postulated small RNA was the adapter.

Look at the adapter hypothesis this way. You are a teacher with a room full of small children (amino acids) whom you want to seat at rows of desks alphabetically. You solve the problem by equipping each desk with a lock and giving each child a key. The child whose name begins with A gets the key to desk number 1, and so on. You tell the children that their key will fit only one desk, and that desk is to be their regular location. Thus, children (amino acids) are arrayed in an unambiguous sequence of children (a protein) at a sequence of desks with locks (template) using keys (adapter RNA)!

That adapter idea was a brilliant intuitive leap, for at the time there was no direct evidence for an RNA template and certainly no evidence for the existence of anything like the postulated adapters. Crick did not publish his idea, but he did write it down as a "note to the RNA Tie Club"* in 1955 and discussed it at meetings in 1956.

Zamecnik and I, unaware of Crick's idea, were working just at this time (late 1955 and 1956) on the complex system that would now make protein in the test tube without benefit of living cells. It consisted of ribosomes, a complex mixture of other cell parts containing the enzymes that ac-

---

*A group of 20 scientists interested in sequencing and coding problems. Each member received a necktie embossed with the name of one of the 20 amino acids. There were four honorary members, each with a necktie showing one of the nucleotide bases of DNA.

tivate amino acids, and a supply of ATP and radioactive amino acids.

Zamecnik wondered whether various cell parts might be making RNA as well as protein, so he incubated samples of the protein-making system with radioactive nucleotides to see whether the nucleotides went into RNA. As a test to be sure radioactive amino acids would *not* be put into RNA — a kind of control experiment — he added some more tubes, identical to the first but with radioactive amino acids instead of radioactive nucleotides. Then he extracted and counted RNA from all the tubes. To our complete surprise, the RNA in *the tubes incubated with radioactive amino acids became radioactive.* Amino acids had somehow got attached to RNA.

Zamecnik and I pursued this unexpected finding with growing excitement. We showed that different amino acids became chemically linked to hitherto unknown, relatively small RNA molecules. The enzymes that accomplished this feat were the same ones that activated the amino acids initially, the ones I had discovered the year before. These had to be remarkably versatile enzymes, able to activate amino acids and then quickly transfer them, still activated, to a special RNA. Furthermore, when we isolated these RNA molecules with attached amino acids and put them back with ribosomes, all of the amino acids were promptly turned into protein! We had quite unexpectedly stumbled upon something very much like Crick's adapter RNAs! His dream was our reality. The new RNAs came to be known as *transfer RNAs.*

So it was that by 1957 we had a well-defined system that could make protein: amino acids; the enzymes to activate them using the energy in ATP; an assembly site (ribosome); and adapters that made it possible to locate precisely the amino acids on a template. But we had no template! The only available candidate for the role of template was the resident RNA of the ribosomes, as we've said. However, there was simply no evidence that ribosomal RNA was the message needed from DNA to order amino acids. There was an uncomfortable feeling about this among workers in the field. Usually, when so many of the parts of a puzzle are in place, the final ones come more easily. Something was missing.

The answer, again, was unexpected. The discovery of transfer RNA drew Crick and me together. As a consequence, I spent the academic year 1957–58 at the Cavendish Laboratories in Cambridge. That was a year of peak excitement in molecular biology, and the Cavendish was humming with heated discussion and many visiting scientists. In January of 1958 I visited the Institut Pasteur in Paris and first came upon the germinating ideas and experiments that were very soon to solve the template mystery. There, Jacques Monod and François Jacob had about a year earlier begun to merge their talents to tackle problems of gene expression and control. When I arrived, they and a colleague visiting from the United States, Arthur Pardee, had just begun a series of experiments which came to be known as the PaJaMo experiments (for the first two letters of the names of the participants). These famous experiments perturbed the status quo of protein synthesis in particular and molecular biology in general, and opened a bright new vista.

When bacteria mate, as you'll recall, a connection between male and female is formed, through which the male injects its DNA at a steady, predictable rate. Jacob had contributed substantially, remember, to the understanding of this bacterial conjugation system, and particularly to its use to map the bacterial chromosome. Monod had delineated one of the important gene systems in bacteria, *the β-galactosidase system,* and it was the exploration of this important system that produced the new perturbation. The system consists of the genes for two enzymes: *β-galactosidase,* which makes it possible for a bacterium to digest certain sugars, and *permease,* a membrane protein that allows a bacterium to transport these sugars through its cell membrane, from outside to inside the cell. These two functions are obviously related: the bacterium cannot digest a sugar until it has brought it into the cell.

Here are the relevant parts of the PaJaMo experiment. Some preliminary experiments showed that when a male carrying a normal β-galactosidase gene mates with a female that can't make the enzyme because of a damaged β-galactosidase gene, the female starts making enzyme almost immediately after the entry of the male's normal healthy gene. (Remember, the time of entry can be predicted quite precisely with Jacob's interrupted mating technique.) This seemed to mean that essentially all the machinery for making a new protein was present in a cell *before* the "knowledge" or information for making the protein was presented to it!

Although the Institut Pasteur group had no basic quarrel with the scheme of protein synthesis worked out in the Huntington and other labs, everyone was puzzled by this result. It seemed to say that a cell was all equipped to make a particular protein, and only needed a simple triggering signal to start. Yet that signal had to contain the necessary information! It didn't seem to need more machinery, that is, more ribosomes. The experiment did not offer hard evidence, but it set all of us to wondering: How could a complete assembly plant suddenly start following a new set of blueprints without some substantial tooling up? The female had started right off making new protein immediately on receiving new DNA instructions. DNA was the *only* thing the male had contributed. There was much banter during that winter visit to the effect that the whole activation–transfer RNA–ribosome scheme so thoroughly worked out by Zamecnik and me and others was an experimental trick played by rat livers, and that protein was really made directly on DNA! I left Paris soon after and returned to Cambridge, where there was little interest in PaJaMo; by summer, I was back in Boston.

In late 1958 one might say there was a malaise in the now merged informationist-structuralist-biochemist molecular biology community. The problems that lay at the very heart of information transfer remained unsolved despite so much progress. We had presented science with a good protein synthesis machine that lacked an obvious way of getting the essential information from DNA. Evidence was coming in fast by this time that bacteria used essentially the same machinery as animal cells. The widely held assumption of colinearity between DNA and protein was just as attractive but just as unsubstantiated.

And then all the pieces began to fall into place. The solution was based on the PaJaMo experiment, the combined intellects of Jacob and Monod, Brenner and Crick, and the memory of earlier, forgotten experiments. In 1953, Hershey had noticed that immediately after phages injected their DNA into their bacterial victims, a small amount of RNA was made very rapidly in the bacterium. Because it was made very fast and in very small quantity in response to the entry of phage DNA, it seemed clearly not to be ribosomal RNA. Hershey didn't see what to do with the finding, published it, and forgot about it.

Three years later, Elliot Volkin and Lazarus Astrachan, at Oak Ridge National Laboratory in Tennessee, had made similar observations, adding the enticing morsel that the nucleotide composition of the small, rapidly made RNA resembled that *of the infecting phage's DNA*. It appeared that the phage was causing to be made inside the bacteria a new RNA that resembled the phage's own genetic material! Volkin and Astrachan, like Hershey, had nowhere to plug in the finding, so they, too, put it on the shelf.

We cannot but wonder at the accelerated pace of biology. Gregor Mendel's and Archibald Garrod's experiments could be forgotten for decades. Here two lesser experiments were brought out of moth balls only a few years after they'd been retired.

Because the phage's single-minded purpose is to sabotage the bacterium's protein-synthesizing machinery by converting it to the manufacture of phage protein, it was quickly guessed by Jacob, Crick, and Brenner at a meeting one Sunday in 1959 in Cambridge that the RNA was carrying a message from phage DNA to bacterial ribosomes. Could it be that in the PaJaMo system the entering β-galactosidase gene immediately started making just such an RNA? Might the male β-galactosidase gene produce a template RNA that would become attached to the female's ribosomes and so cause them to hook up amino acids in the proper order to make the enzyme β-galactosidase? Such an RNA would be the long-awaited template: a copy of a gene's length of DNA, to be read by the ribosome–transfer RNA reading machine into the gene's sequence specification.

That inspired interpretation was precisely on the mark. Within months, Brenner and Jacob had solicited the help of Meselson at Cal Tech, an expert in physical characterization of RNA and DNA, to do a sharp test of the theory.

That experiment showed what Brenner, Jacob, and Meselson fervently expected it would show. Using the phage system, they found that the new template RNA that appears in phage-infected bacteria *does* bind to the ribosomes of bacteria. It was one step only in the final proof. It was the first direct demonstration that the postulated template RNA had the properties expected of a template. In the coming few years, proofs of the validity of the concept came in rapidly.

In the fall of 1960, Jacob and Monod christened the new species of RNA *messenger RNA*, for it carried the message from gene to protein synthesis machinery.

CHAPTER TWELVE

The Code Is Cracked

Steps in protein assembly. Francis Crick and Sydney Brenner tackle the code with genetics. Marshall Nirenberg and Johann Matthaei crack the code with biochemistry.

THE ASSEMBLY PLANT for protein was pretty well understood by the spring of 1961. There remained one great unknown whose solution should punctuate the end of the first bright chapter of molecular biology: the language of DNA, the code. We'd worked out the mechanical and electrical features of a wireless transmitter, but didn't know what the dots and dashes meant!

Let's review the steps in protein assembly, the wireless machinery. The sequence of nucleotides in one of the strands of DNA corresponding to a gene — the information for the ordering of amino acids in one continuous protein chain — is copied by an enzyme called RNA polymerase into a single strand of RNA. That single strand is called messenger RNA. The enzyme is very similar to Kornberg's DNA polymerase, which, as you remember, uses triphosphates of the nucleotides of DNA to make new DNA strands on a template DNA strand. RNA polymerase uses triphosphates of the nucleotides of RNA to make new RNA strands on a template DNA strand. Each messenger RNA is released from its template DNA strand and becomes attached to a ribosome, thereby programming the ribosome

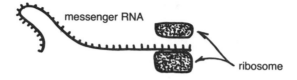

to make a specific protein. The ribosome provides a physical matrix on which messenger RNA's long tape-like single-strand sequence of nucleotides can be read.

The amino acids to be ordered and coupled arrive at the ribosome reading site attached to transfer RNA molecules (Chapter XI). Each amino acid is attached to its own special transfer RNA (a child with a desk key in our school room). This is accomplished by the amino acid's own activating enzyme. Thus, each amino acid is unmistakably identified as it arrives at the messenger. The messenger presents to the transfer RNA sets of three nucleotides (called *codons,* the lock) to which the transfer RNAs match their *anticodons* (the keys).

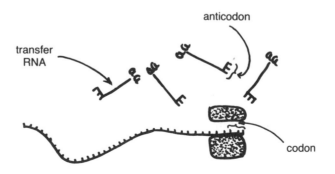

Codon-anticodon matching (the key entering the lock) is accomplished by hydrogen-bonded pairing of the codon-anticodon triplets in the standard manner of interaction between complementary segments of RNA or DNA.

The ribosome, holding the messenger RNA in reading position, exposes the first messenger codon. The transfer RNA with the right anticodon is selected from the many transfer RNAs crowding about and is locked in place on the first codon. The second codon next selects the anticodon of the second transfer RNA. As soon as *two* transfer RNA–amino acids are properly located on the messenger RNA at contiguous codons, the two now nearly touching amino acids can (because they remain activated) link up with one another. And so a protein chain begins.

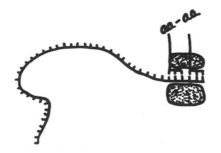

The transfer RNA that entered first now departs, and the two linked amino acids remain attached to the messenger via the transfer RNA that entered second.

Now the ribosome hikes the messenger along three nucleotides. This motion of the messenger on the ribosome requires
energy, of course, and the energy is supplied by another

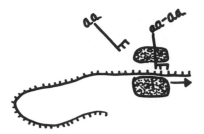

ATP-like molecule called GTP (guanosine triphosphate).
The move readies the reading site for the *next* transfer
RNA–amino acid, and so completes a single linking cycle.

*

The process is now simply repeated in the same series of
steps: a third transfer RNA is selected; its anticodon binds
to the codon on the messenger now in place; its amino acid
links up to the second of the previously linked amino acids;
the second transfer RNA departs; and a protein three amino
acids in length is bound to the reading machine by the last-
entered transfer RNA. The GTP-energized triplet hike is
repeated, and the cycle is ready again. Note that the chain
grows from one end to the other and is always held to the

messenger-ribosome by the last-entered transfer RNA. As synthesis progresses, the ribosome, which looks something like an acorn, would have threads too fine to see in a microscope emerging from it: the spent, already read forward end of the messenger RNA, the to-be-read tail, and the growing protein chain:

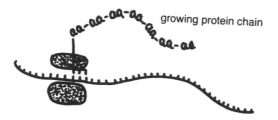

growing protein chain

The average protein chain is about 300 amino acids long. When the last amino acid has been added, a special enzyme releases the chain from the last transfer RNA. The new chain then completes its spontaneous folding into its native shape, a process prescribed by the number and kind of amino acids in the chain.

The process of copying DNA into RNA is called *transcription*. The process of reading RNA on ribosomes to make protein is called *translation*. The names reflect the obvious analogy with language.

Of all the cellular processes that science has discovered, this is the one that most compels my awe. Here, information — a set of instructions written out in four letters — *becomes* living substance. Life's plan becomes life's reality.

❋

Many of the best minds of biology were now preoccupied to varying degrees with how to get at the code. After the discovery of transfer RNA, there was a flurry of excitement that the code might be revealed by trying to discover the triplet of bases (the anticodon) in transfer RNA that corresponded to the particular amino acid attached to it. Because the anticodon was a base sequence with which only one amino acid was unambiguously associated, there might be an opening there. But when it became apparent that each transfer RNA was about 70 nucleotides long and that there were no rules for recognizing an anticodon, this hope died. Another way to get at the code, the most promising at the time, was an extension of the hemoglobin and colinearity approaches: produce mutations in DNA, map them accurately, determine the corresponding colinear amino acid changes in protein, and try to deduce codons. Fine-structure genetics, plus a refined knowledge of what kind of base changes mutagens produce, made such an approach possible. It would have been a long and arduous task. (As it turned out, this approach was most helpful in confirming the code.) A third approach was the most obvious: a dogged analysis, nucleotide by nucleotide, of the complete nucleotide sequence of the messenger RNA for a protein whose full amino acid sequence was known. An imperfect technology could have done this, but it would take a long time. The prospects seemed grim in the spring of 1961 for any early solution. It would take another five or ten years of hard work, most scientists thought.

Facts about the code were scarce, but ideas were plentiful and had been percolating for some years before the discovery of messenger RNA. Some had blossomed and then wilted for lack of experimental nourishment; they had been premature.

It was generally true in the two decades of molecular biology that the important ideas were well ahead of experimental evidence supporting them. Thus, the sequence hypothesis, the central dogma, and ideas about coding were widely accepted long before they were established as facts. This is fairly common in science, but in this case, unlike most, the ideas were right.

Messenger RNA was the turning point. With its confirmation as the link between DNA and the protein-assembly system, coding speculation got a firm footing. Francis Crick and Sydney Brenner, more than any others, penetrated the mystery. Their thinking went as follows.

Remember Chargaff's finding that the nucleotide composition of DNAs from all sorts of different animals, plants, and bacteria varied greatly? The nucleotide composition of ribosomal RNA of all sorts of different animals did not show such variation; it was quite uniform. If ribosomal RNA were a template, it should be more like DNA. Messenger RNA, on the other hand, was clearly the DNA copy. Its nucleotide composition was similar to DNA's, varying in different species as DNA varied. However, the amino acid composition of the total protein of cells varied little from species to species. So one was forced to conclude that each amino acid must be represented by more than one — probably several — triplet nucleotides, that is, codons. The word for that was *degeneracy;* the code was degenerate. If many more than 20 codons, representing 20 amino acids, are strung out end to end in a 1,000-nucleotide message, how can they be read unambiguously? Only by starting to read the message at one fixed starting point and reading through to the end, triplet by triplet. Thus Brenner and Crick anticipated degeneracy of the code and the need for sequential reading from a fixed point.

In 1961, Crick and Brenner used a set of truly inspired experiments to address the idea of reading from a fixed point by triplets. It was the last major contribution of microbial genetics to the information puzzle. (The results of the colinearity studies were yet to come in; they arrived between 1962 and 1965. The ongoing work was familiar to those in the field and the outcome was taken for granted.) It was fitting that Crick and Brenner, who had contributed so prodigiously to the advance of coding theory, should produce this ingenious experimental finale.

Brenner and associates Alice Orgel and Leslie Barnett had for some time been studying mutations produced in phages by acridine dyes. Their discoveries led them to theorize that these particular dyes acted on DNA by causing addition or deletion of single nucleotides.

start: $\overline{ABC}$ $\overline{ABC}$ $\overline{ABC}$ $\overline{ABC}$ $\overline{ABC}$ $\overline{ABC}$ $\overline{ABC}$

We picture here the bases of DNA and messenger as a repeating sequence, ABC. This makes it easier to show when something goes wrong. The brackets are the presumed reading frame (the byte size of the reading machine).

Brenner and Crick got the idea that if messenger RNA had an addition or deletion of a nucleotide somewhere along its length, and if the messenger RNA were read from one end to the other continuously, the message would be completely misread beyond the absent (or added) nucleotide, thus

The effect of a nucleotide *deletion*

—————— gibberish ——————→

start: $\overline{ABC}$ $\overline{ABC}$ $\overline{ABC}$ $\overline{BCA}$ $\overline{BCA}$ $\overline{BCA}$ $\overline{BCA}$ $\overline{BCA}$

↑

A removed

(A nucleotide *addition* would have the same effect.)

Messenger RNA is a direct copy of DNA, so any mutation made in the DNA by acridine dyes would be copied into it. The situation is entirely analogous to one in which a book is read by a machine that processes letters mechanically in groups of three. A deletion or addition of one letter will make the reading beyond the error total gibberish.

Brenner and Crick further predicted that if a phage's DNA containing a nucleotide *deletion* were recombined with another phage's DNA containing a nucleotide *addition,* the reading would be corrected and normal function would be restored beyond the second change.

—— normal ——→

start: $\overline{ABC}$ $\overline{ABC}$ $\overline{ABC}$ $\overline{CAB}$ $\overline{CAB}$ $\overline{CAC}$ $\overline{ABC}$ $\overline{ABC}$ $\overline{ABC}$

	↑		↑		
	C		B		
	added		removed		

Their third prediction was that if a phage's DNA containing a nucleotide deletion were recombined with another phage's DNA containing *another* nucleotide deletion, the combined deletions would *not* be corrective, that is, would not restore function. The same would be true for two additions. Reading would be expected to remain out of register.

—— gibberish —— still gibberish ——→

start: $\overline{ABC}$ $\overline{ABC}$ $\overline{ACA}$ $\overline{BCA}$ $\overline{BCA}$ $\overline{CAB}$ $\overline{CAB}$ $\overline{CAB}$ $\overline{CAB}$

	↑		↑		
	B		B		
	removed		removed		

Finally, they expected that DNA made by recombining DNAs from *three* phages, all having additions or all having deletions, *would* restore normal function and produce messenger RNA that could be read normally beyond the third change.

```
                              ——— normal ———→
start: ABC  ABC  AAB  CAB  CCA  BCA  ABC  ABC  ABC  ABC
            ↑         ↑    ↑
            A         C    A
          added     added added
```

These various combinations of nucleotide deletions and additions were constructed by causing recombination to occur in living phages. The phages carrying them were examined to see whether they were defective or were restored to normal function. In all cases, the predictions were borne out!

In the above case of recombined phage mutants carrying a nucleotide deletion and, down the line, a nucleotide addition (or vice versa), it was expected that the protein produced by the altered gene would contain a sequence of unfamiliar amino acids — a meaningless sequence — between the two mutations.

At the time, it was not possible to obtain the relevant protein. However, a few years later it was isolated by other researchers and found to contain the expected meaningless amino acid sequence! The outcome vindicated Brenner's assumptions about the mode of action of acridine dyes, which have since become valuable experimental tools in genetics. It brought indirect but very strong evidence for a triplet code and further showed, again indirectly, that the message is read from a fixed starting point.

❊

But the code still lay hidden. While Crick and Brenner were beginning the experiments just related, in the spring of

1961, the charge that was to crack the coding seam wide open was being rammed into place at the National Institutes of Health in Bethesda, Maryland. Marshall Nirenberg and Johann Matthaei were young biochemists who were just starting a study of protein synthesis. They wanted to find some direct evidence for the existence of a template in a system making protein.

There was nothing particularly unusual about the system they used. It was derived from broken-up bacteria and consisted of ribosomes, transfer RNAs, ATP, GTP, and the necessary enzymes. They had *several* radioactive amino acids present in all test systems. They removed RNAs from ribosomes, viruses, and other sources and asked whether these RNAs would boost the amount of protein synthesis that occurred. Their best guess, as was most everyone's at the time, was that the template would be associated with ribosomes; possibly, it would be ribosomal RNA itself, but hopefully something that could be removed from ribosomes.

You will remember from the last chapter that in the fall of 1959 the Cambridge and Paris groups had put together the messenger RNA concept. In the spring of 1960, Brenner, Jacob, and Meselson had provided the first supporting evidence for the concept. But nothing had yet been published and researchers who were not members of the intimate molecular biology fraternity would not necessarily have learned of the new developments. So it was that Nirenberg and Matthaei were innocent of knowledge that might have discouraged them from proceeding with their experiments.

They found that viral RNA was a good stimulator of protein synthesis; other natural RNAs caused some small stimulation but didn't seem very promising. In addition to the natural RNAs they were trying out in the system, they also tried certain artificial RNAs. These were being made, by chance, in a neighboring lab. Artificial RNAs were put together by a newly discovered enzyme and could be made to contain one, two, three, or four nucleotides in any amounts desired. One of these artificial RNAs was polyuridylic acid (polyU), a long sequence of nothing but the nucleotide of uracil linked together exactly as nucleotides are in natural RNA. When they added this polyU to their protein synthesis system, there was an enormous burst of incorporation of radioactive amino acid into something that seemed to be protein!

Nirenberg and Matthaei, amazed, quickly sought to find out which radioactive amino acid(s) was being so exuberantly thrust into "protein" with the help of polyU. For, you will recall, they added a mixture of radioactive amino acids to the system. They found only one, phenylalanine, and the "protein" made was polyphenylalanine, that is, phenylalanine repeating itself over and over in a long chain!

Of all the experiments performed in the fifteen years of molecular biology's train of successes, this one certainly deserves the highest rating for sheer surprise. Here was complex normal biological machinery using an artificial messenger to make an artificial protein. This was like providing an animal with synthetic sperms and eggs and producing a robot! But with surprise came another astonishing realization: if the cell's natural machinery could be made to make any kind of protein by giving it any kind of RNA, why wasn't this the answer to the code? If coding theory was right, then surely the codon for phenylalanine was now known — U-U-U!

Nirenberg reported his and Matthaei's now legendary experiment at the Fifth International Congress of Biochemistry in Moscow in August of 1961. Meselson, Benzer, Crick,

Watson, and Perutz were there to hear. (Most of the Russians didn't have the vaguest idea what was going on, because they were still under Lysenko's blanket suppression of genetics, Darwin, and molecular biology.) The Nirenberg paper was the high point of the meeting, and those in the know were convinced that the code was on the verge of being revealed.

They were right! Each year from 1961 until 1966, code triplets (codons) for the 20 amino acids were discovered and added to biology's vocabulary. Nirenberg's rapidly expanding lab was in the forefront of the effort, closely followed by the lab of Severo Ochoa, Professor of Biochemistry at New York University, and discoverer of the method for making the artificial RNAs. This was one of the tensest competitions science had ever known. Nirenberg was to receive the Nobel Prize in 1969. Ochoa had already got the Prize in 1959 for discovering the enzyme that made the synthetic polymers.

Let's have a look at the methods they and later others used to discover the codons for the 20 amino acids.

The standard system is that used by Nirenberg and Matthaei to reveal that U-U-U was the codon for phenylalanine. Assuming that the code is based on triplets, the triplet U-U-U is assigned to phenylalanine. The object now is to add new polymers, containing the nucleotide U and other nucleotides as well, to see which amino acids are stimulated to be incorporated into protein.

The nucleotides in these artificial polymers are linked to each other exactly as nucleotides are in natural RNA, as we've said. But the distribution of the four different nucleotides along the chain is completely random. So the probability of the occurrence of any particular triplet is a matter of pure chance, influenced only by the number and kind of nucleotides in the polymer. Thus, for example, the probabilities of the particular triplets U-U-U, U-U-C, U-C-C, and CCC occurring in a polymer containing only U and C (polyUC) can be calculated. For example, in a polyUC that contains five times as much U as C, if the relative frequency of U-U-U is set at 1, the chance of occurrence of the triplet UUC would be $1 \times 1 \times 1/5 = 0.2$; the chance of occurrence of UCC would be $1 \times 1/5 \times 1/5 = 0.04$; and of CCC, $1/5 \times 1/5 \times 1/5 = 0.008$.

Twenty identical tubes containing the natural protein-making machinery are lined up, and the polymer is added to each. *All* the tubes have *all* 20 amino acids in them, but in each tube a different one is labeled. The only difference between tubes is the amino acid that is radioactive. Now the tubes are incubated. At the end of the incubation, the protein formed in each tube is isolated, and its radioactivity is measured. Because the polymer contains considerable U, the artificial protein formed will contain much phenylalanine, whose relative incorporation is arbitrarily called 1. The extent of incorporation of each of the other amino acids is then noted in relation to that of phenylalanine. These figures are compared to the probability of occurrence of a particular triplet as calculated above, and where good matching is obtained, a tentative triplet assignment is made.

Here's an example: polyUC containing 39 percent U and 61 percent C was used. The relative frequency of occurrence of UUU in the polymer is called 1; the probability of occurrence of UUC is calculated to be 1.57 ($1 \times 1 \times 61/39$); of UCC is 2.44 ($1 \times 61/39 \times 61/39$); and of CCC is 3.82 ($61/39 \times 61/39 \times 61/39$). The experiment is done and the incorporation into protein of several amino acids is found to be: phenylalanine 1, arginine 0, alanine 0, serine 1.60, proline 2.85, tyrosine 0, valine 0, and so on.

Serine was incorporated into the protein 1.60 times more than phenylalanine, corresponding very closely to the frequency of the triplet UCC (1.57) in the polyUC messenger. So the triplet UUC is assigned to serine.

Proline entered protein 2.85 times more than phenylalanine, and the triplet UCC occurred in the messenger in amount closest to that (2.44). So the triplet UCC is assigned tentatively to proline. This technique allows assignment of codons to most of the amino acids. Of course the *order* of the nucleotides in the codon is not known.

A second technique, also developed by Nirenberg and his associates in 1964, depends on the surprising finding that under certain conditions, chemically made nucleotide *triplets* of *known* sequence will bind readily to ribosomes and, in turn, will promote the binding of specific transfer RNA–amino acids. This kind of experiment not only most satisfactorily supports the triplet code, but also permits assignment of the *order* of bases in the triplet.

A third approach was pioneered by Har Gobind Khorana, a wizard at chemically creating messenger RNAs of considerable length and predetermined sequence. These man-made messengers were added to the natural protein-making system and caused new protein to be made. In this case, the amino acid sequence of new protein could be compared directly with the nucleotide sequence in the messenger that guided its synthesis.

Still, a small fear lurked behind all this work. Protein-building machinery taken out of cells seemed not to care that it was given fake messages. It went right ahead and made fake protein. Perhaps the whole thing was fake, not doing what it did in real life. But too many predictions about the code had already been borne out: the triplet nature of the code, degeneracy, and continuous reading from a fixed starting point. And the argument could be turned around: the very fact that it was nature's own system that was making protein in response to any message was profoundly convincing. However, proof that the code was nature's own true code awaited some correlation of these outside-the-cell experiments with events in the living cell. This came when mutagenic agents that produced specific, known base changes in DNA in live organisms caused amino acid changes in protein that could be predicted from the new codon assignments.

Here's the way the code is usually presented, as designed by Francis Crick:

1st ↓	2nd → U	C	A	G	3rd ↓
U	Phenylalanine	Serine	Tyrosine	Cystine	U
	Phenylalanine	Serine	Tyrosine	Cystine	C
	Leucine	Serine	Nonsense	Nonsense	A
	Leucine	Serine	Nonsense	Tryptophan	G
C	Leucine	Proline	Histamine	Arginine	U
	Leucine	Proline	Histamine	Arginine	C
	Leucine	Proline	Glutamine	Arginine	A
	Leucine	Proline	Glutamine	Arginine	G
A	Isoleucine	Threonine	Asparagine	Serine	U
	Isoleucine	Threonine	Asparagine	Serine	C
	Isoleucine	Threonine	Lysine	Arginine	A
	Methionine	Threonine	Lysine	Arginine	G
G	Valine	Alanine	Aspartic acid	Glycine	U
	Valine	Alanine	Aspartic acid	Glycine	C
	Valine	Alanine	Glutamic acid	Glycine	A
	Valine	Alanine	Glutamic acid	Glycine	G

The bases of the four nucleotides of RNA are designated as U, C, A, and G. The first base of every triplet is in the left column; the second base, across the top; the third, down the right side. Thus, for example, the codons for the amino acid phenylalanine are UUU and UUC; those for the amino acid arginine CGU, CGC, CGA, CGG and AGA, AGG. Three triplets UAG (nonsense 1), UAA (nonsense 2), and UGA (nonsense 3) specify no amino acid and are used to designate the end of a chain. When a mutation causes such a triplet to appear, the protein being made is terminated prematurely (see Chapter X).

One fascinating feature of the code is its *degeneracy*, as predicted by Brenner and Crick. Degeneracy, remember, means that most of the amino acids have more than one codon. Thus, as the table shows, three of the twenty amino acids have six codons; five have four codons; one has three; nine have two; and only two amino acids have one codon.

The second, very dramatic aspect of the code is its *universality*. All studies have revealed that synthetic machinery and messengers are interchangeable between species as different as bacteria, plants, and mammals. Thus, the code is the same for all living creatures. The fact that all living creatures from the lowliest to humans use the same 20 amino acids, the same four nucleotides, the same code, and pretty much the same protein synthesis machinery, is the strongest possible confirmation for our common evolutionary origin.

CHAPTER THIRTEEN

Science in Our Lives

Science's discoveries enrich our lives philosophically and aesthetically as well as practically. Darwinian natural selection and the intimate relatedness of all living forms are confirmed by molecular biology. The ways of science are unique in revealing verifiable truth. New realms of exploration stretch before us since the cracking of the code.

I REMEMBER that among our many courses in medical school, Anatomy was one we students looked upon with particular disdain. The course required that we memorize endless lists of parts when we might otherwise have been immersed in the loftier intellectual challenges of physiology and biochemistry. Yet Anatomy, more than any other course, has stuck with me over the years. Why? I suppose it is because I live intimately with those very tangible parts whose names are now inscribed in my brain. Those muscles and bones and liver are *me*. Anatomy was my introduction to my biological self, and it made me more comfortable with my own machinery.

Knowledge of the elements of form and function of *all* living cells; of the basic principles governing the operation and propagation of life; and of how living uniqueness evolved and will be passed on into the future have value for us, scientist and nonscientist alike, in the aesthetic and practical enrichment of our lives. The discoveries of science hold out to us the opportunity to align our lives in closer harmony with the insights into nature they provide.

Consider evolution. The biological data collected by Charles Darwin on his famous voyage of scientific discovery on the *Beagle*, together with an additional large body of supporting evidence accumulated in the last century, had, by the 1940s, put evolution on a firm base. But definitive verification of Darwinian origins in the surer terms of the chemistry and physics of life processes awaited the revelations of molecular biology.

In Chapter III, we saw evidence that the two mechanisms underlying evolutionary change — chance mutation of DNA and environmental selection of a better adapted variant — can be reproduced in detail in the life of populations of microorganisms. This landmark discovery brought the earth's most abundant form of life, bacteria, into the evolutionary fold. This made it possible to observe evolution in the laboratory and to study it in detail in organisms that reproduce 100,000 times faster than we do, and are themselves among the most ancient on our planet.

The discoveries of molecular biology have solidly established the Darwinian view of evolution. The central dogma of molecular biology — a term coined by Francis Crick — states that the instructions for influencing living processes flow in only one direction, from DNA to RNA to protein. DNA is the source of heritable information. Its information is translated into protein via RNA, and protein, the material of which all living beings are made, cannot pass information back to, or influence, DNA. It is a one-way process.

Protein is the final repository of DNA's information until that information, in a given organism, is destroyed by the decay that follows death.

The vision of being able to influence or modify evolution by outside manipulation à la Lamarck (Chapter III) is thus finally extinguished by the discoveries of molecular biology.

The intimate relatedness of all living creatures from bacteria to plants to animals is conclusively demonstrated by the facts that all our proteins are made of the same 20 amino acids, all our DNAs and RNAs are made of the same four bases, and we all use the same genetic code and the same machinery for translating the instructions of DNA into protein. Even the widely shared mechanisms for generating energy for life processes are quite similar in all of us.

In Chapter IX we learned of the method of mixing denatured DNA molecules from different animals and allowing them to reform as hybrid double strands. This technique has become an extremely sensitive tool for assessing relatedness of, or differences between, species. It has proved to be much more accurate than more conventional anatomical methods. And the results of these new DNA hybridization techniques have gratifyingly confirmed species relatedness and presumed lines of evolutionary descent obtained by other, more conventional, methods. Thus, cellular chemistry provides solid confirmatory evidence for evolution.

The core of genetic and evolutionary phenomena, then, includes the *unidirectional flow of information,* the possibility of altering the information by *mutation,* and the *selection* of variants so produced. This core by no means excludes the possibility of an organism's losing or gaining significantly larger pieces of genetic information from extraneous sources, with consequent modification of offspring. The most useful of such sources is sex. The rearrangement of parental genes in the chromosomes before formation of sperms and eggs, and the subsequent incorporation of the novel DNAs into the cells of offspring, insures substantial input of new genes and new combinations of genes at each generation. Sex must have appeared very early in evolution, because mutation alone is slow and chancy.

There are other prevalent but less obvious mechanisms of gene transfer. Viruses, which afflict bacteria, plants, and animals, have the outrageous capacity to enter the cells of their hosts, pick up genes from their hosts, knit them by recombination into their own DNA, and carry them to another bacterium or animal host. Viruses thus act as gene transfer agents. There are also viruses, including those that cause cancer in animals and birds, whose genes are made of RNA rather than DNA. When these viruses attack host cells, they cause DNA to be made in the image of their RNA genes, the reverse of part of the central dogma (but not of the irreversibility of translation into protein). The DNA, integrated among the host genes, can then direct the synthesis of more viral RNA and, hence, more viruses. In the process, host genes may also be copied into viral RNA and thereby be transmissible to new hosts. The possibility that whole genes or groups of genes could be gained by such mechanisms during evolution is very real.

The most widely held view of life's beginning is that simple molecules present in the nourishing seas of our earth progressed slowly to a self-replicating cell over vast stretches of time beginning 3 or 4 billion years ago. A less widely held view is that life on our earth might have been deposited, in one or another primordial form, from outer space, perhaps riding in on an impacting comet. This second view simply displaces the problem of the origin of life to another celestial body.

Regardless of the size of the quantum jumps in information at the beginning of life and during evolution, every living creature is the result of an enormous number of chance events. It follows that if we were to wipe the evolutionary slate clean and start the whole process over again, using the same rules, the chance of reproducing any creature we know today, including humans, would be vanishingly small. That's because the chance of repeating any long series of chance events is so small as to be essentially nil. We are thus not the end of some preset plan, and we are not governed by rules different from those governing all other creatures. Our uniqueness is the pragmatic uniqueness each and every creature shares on this earth: we can never be made again. *Our* very special uniqueness lies in our possession of evolution's most exquisite development, the human brain — an organ that has turned around and discovered these truths about our origin!

*

The scientist undertakes work in the faith that life, the world, the universe, are orderly and are based on the principles of chemistry and physics. The discovery of the molecular basis of life as described in the previous chapters and the demonstration of its adherence to rational laws is a deeply satisfying validation of science's ways. We still have a great deal to learn, but we can at least be virtually certain that new knowledge will be interpretable in the language and principles that science has already established.

Science owes its success to the conditions it imposes on itself to substantiate the claim that something is *true*. The scientist is able to build steadily truth upon truth through adherence to rules of procedure that require truths to be verified. Mystics, vitalists, theists, cultists, and astrologers also seek and proclaim their own "truths." But these are significantly different from the truths of science, in that they must invoke inexplicable, enigmatic powers and forces that are untestable and unverifiable, and so remain safe from challenge. The strength of science is in its admission of its vulnerability, its unwillingness to trust its own conclusions until they have been tested by others.

In past centuries the institutions of government and organized religion often used their power to suppress scientific knowledge and subject scientists to imprisonment, torture, and death. It was only some 600 years ago that a monk, Bruno, was burned at the stake for teaching that the earth was round. That the suppression of science is no longer the case is due to science's steady construction of an overwhelmingly convincing edifice of enlightenment, from which has flowed an abundance of practical benefits. Authoritarianism and intolerance have been mitigated by both the methods and the discoveries of science. Discoveries that were once heretical have become comfortable verities. The questioning of nature, once feared, is no longer seen as threatening. Science has thus had a significant role in humanizing our society.

❋

The passion for exploration is deeply rooted in all of us. Indeed, many of the great forward steps in the history of humankind have in one way or another been related to exploration, whether on the oceans, the land, in space, or within the microworld explored by physics and biology. The ways of basic exploratory science are refreshing in their relative simplicity, openness, knowledge-sharing, and reverence for truth. The discoveries we have discussed in this book, with their enormous potential for influencing human welfare, were performed openly, discussed widely, repeated, and extended as the work progressed.

Avery's demonstration that genes are made of DNA was immediately exploited by Chargaff on the one hand and Lederberg on the other. Watson and Crick soon thereafter plugged Chargaff's Avery-inspired data into the evolving structure of DNA. Lederberg's work directly inspired Hayes' elucidation of the sexual process in bacteria. Hayes, in turn, set Wollman and Jacob on the trail of a whole new method of mapping genes that led to the PaJaMo experiment and its dramatic consequences. After the discovery of transfer RNA, even before publishing our work, Zamecnik and I felt on our necks the hot breath of colleagues eager to learn techniques that would enable them to enter the field, an experience almost all scientists have had. Gone is the day when the work of a Garrod or a Mendel could go unexploited for decades.

The phage group, by openly discussing and sharing ideas, stimulated a new way of looking at problems. Francis Crick was a kind of nerve center in Cambridge with whom scientists from the United States, England, France, and the Soviet Union corresponded openly, receiving information, ideas, and suggesting experiments in return. Other scientists perform a similar function in other areas of research. Scientists generally use the telephone extensively to discuss work with colleagues. They attend meetings and are constantly giving and arranging seminars by visiting scientists. Their applications for grants, the main source of funds to support their research, are reviewed by their professional peers. In the process, the applicant for funds must lay bare cherished research plans before the eyes of colleagues who may be research competitors. An applicant who tries to protect ideas by providing limited information risks having the grant rejected on the grounds of inadequate evidence of competence. Sometimes ideas are stolen, but this occurs infrequently, and the value of the system's openness and relative freedom from politics in evaluation of merit far outweighs the risks. Scientists publish their work, of course, but because articles may take as long as a year to appear in print, they often mail to their colleagues preprints of their discoveries — manuscripts of intended publications — so that their accomplishments will be known as soon and as widely as possible.

Dissemination of knowledge is based on a fundamental pragmatic observation: if ideas, methods, and findings are openly shared, the work of all scientists is advanced and further discovery is facilitated.

The whole implausible process, born of curiosity and sustained by a reverence for truth and enlightened governmental support, has been incredibly effective in opening nature's doors for man. The flow of new knowledge of our world is really only beginning. So many scientists are now at work in the world that, it is estimated, most of the scientists who *ever* lived are alive today.

*

The 20-year surge of discovery described in this book has given only a glimpse of an unexplored universe within ourselves — and within our reach. It has sharpened our curiosity and prodded us to search further, because we have succeeded in uncovering not just phenomena, but generalities, rules, principles, universal mechanisms.

Our story ended with the cracking of the genetic code, but another tale of discovery was unfolding during the same period, a story that would give new insight into the *control of genetic expression*. It was a door to future research opening on areas related to the regulation of change, the uncovering of hidden gene potentials, the acquisition of new genetic properties, as in cancer. The story began with Jacques Monod's fascination with the fact that bacteria have a remarkable capacity to alter themselves in response to changes in the availability of materials they need. Bacteria given food not previously available quickly synthesize an enzyme to assist them in digesting that food. Bacteria making their own amino acids quickly stop making them if these amino acids are presented to them. These are clearly conservation, or adaptation, measures with obvious survival advantage. It should be emphasized that these adaptations reflect built-in capabilities of the bacteria, and are not due to mutation and selection (as, for example, is their acquisition of streptomycin resistance, Chapter VI).

Extensive exploration of these phenomena by Monod, later in company with Jacob, revealed them to be genetically based. The enzymes whose synthesis was turned on or shut off were, of course, governed by genes. It was found that the expression of these genes was under the control of repressor molecules, proteins with the very special role of stopping genes from making messenger RNA molecules. Thus, we came to know that enzymes, which express cell character, can be made to appear or disappear by turning their genes on and off. At least in plastic, adapting, adjusting bacterial populations, there existed a quickly triggered device for controlling the expression of genes already present in the organisms.

These discoveries have had a powerful influence on much of biological exploration since 1960, when they were formally presented by Monod and Jacob. (They received the Nobel Prize in 1965 for this work.) But so far, there is no obvious way to translate our understanding of regulatory phenomena in bacteria to equivalent processes in animal cells. Perhaps this should not be surprising; an animal cell is many hundreds of times more complex than its bacterial ancestor. In addition, the two kinds of organism are faced with entirely different problems. Bacteria multiply or cease growth at the whim of the environment, expressing most of their genetic potential most of the time. Higher cells live in restraint, growing little, under constant internal control and expressing only a small fraction of their total genetic potential throughout their lives. Whatever the level of our ignorance, the problem is ripe for attack. The control of gene expression lies at the root of the most fascinating problem in biology: How is the progression from fertilized egg to emergent organism orchestrated?

*

Molecular biology has continued to generate some fascinating answers, and many more questions. The questions will keep exploration lively for decades to come. We have discovered, for example, that a very large amount of the DNA of higher organisms seems not to be involved in messenger production. It sits in the nucleus with no known function. Is it an undiscarded leftover of evolution? Is it providing information for unknown structures and functions?

Another astonishing recent discovery is that genes of more complex organisms are not single, continuous stretches of DNA as I described in the preceding chapters. They are interrupted! One example is a particular gene that has a coding sequence of 432 nucleotides that codes for a protein of 144 amino acids. But the nucleotide sequence is not continuous; it is interrupted by two sequences of 116 and 646 nucleotides that have nothing to do with the protein to be made. The entire length of the "gene," consisting of the sequence coding for the final protein product and the intervening sequences, is transcribed into a long RNA molecule of $432 + 116 + 646 = 1,194$ bases. Special enzymes clip out the sequences that aren't needed and splice together the final coding sequence to produce a proper messenger RNA, which is then translated on ribosomes. Why all that extra work? We don't know, but we can be certain it has some special significance for the life of the cell.

Important technological advances made during the past decade permit intimate study of gene structure and function. The most dramatic of these is gene splicing, or recombinant DNA technology. This, as we noted at the end of Chapter V, is a method of making large quantities of particular genes that can be used to study cell function and to produce valuable medical and agricultural products. Our ancient microbial ancestors continue to contribute to the advance of knowledge of all forms of life.

Less dramatic but exceptionally useful for providing knowledge of gene structure is the development of rapid and accurate methods for determining the *sequence* of nucleotides in genes. We are also becoming proficient at making genes from scratch by straight chemical means. These new capabilities for producing large quantities of genes and for analyzing their nucleotide sequences are making it possible to gain a much more sophisticated understanding of the gene and, we hope, of the cellular devices for turning them on and off, temporarily or permanently.

The emerging commercial and medical applications of recombinant DNA technology are potent new forces for human benefit to our society. The production of large amounts of medically needed protein products of specific genes by various techniques (insulin and interferon, for example) are obvious examples. Producing products of value in agriculture and improving industrial processes such as fermentations are others. The technology can be applied to agriculture for improving crop yields and for putting photosynthesis to better use to help solve our energy and food supply problems.

*

We live in an age of unprecedentedly vigorous exploration. The scientist is our age's explorer.

The voyage that took [Columbus] to "the Indies" and home was no blind chance, but the creation of his own brain and soul, long studied, carefully planned, repeatedly urged on indifferent princes, and carried through by virtue of his courage, sea knowledge, and indomitable will. No later voyage could ever have such spectacular results; and Columbus's fame would have been more secure had he retired from the sea in 1493. Yet a lofty ambition to explore further, to organize the territories won for Castile, and to complete the circuit of the globe sent him thrice more to America. These voyages, even more than the first, proved him to be the greatest navigator of his age, and enabled him to train the captains and pilots who were to display the banners of Spain off every American cape and island between 50°N and 50°S. The ease with which he anticipated the unknown terrors of the ocean, the skill with which he found his way out and home, again and again, led thousands of men from every western European nation into maritime adventure and exploration. *

History's great land and sea voyages of discovery, aided by science's reassurance that the world was round, made all of the surface of the earth a human domain. As science continues to weave its patterns of knowledge, we reach outward to explore our galaxy and beyond, and inward to learn the reality of ourselves. We no longer need fear or revere mystery, as our forefathers did, unless we choose to. Cumulative knowledge and the means to amass more are at hand to bring us ever deepening insight. The purposes and navigational methods that guide our explorations are not, and have never been, beyond the reach of those who seek to know and have access to learning. A fascinating journey of discovery is in prospect for those whose curiosity, imagination, and sense of adventure compel them to sign on.

*Samuel Eliot Morrison, *Admiral of the Ocean Sea: A Life of Christopher Columbus* (Boston: Atlantic–Little, Brown, 1942).

INDEX

Index